Raspberry Pi 3 Home Automation Projects

Bringing your home to life using Raspberry Pi 3, Arduino, and ESP8266

Shantanu Bhadoria
Ruben Oliva Ramos

BIRMINGHAM - MUMBAI

Raspberry Pi 3 Home Automation Projects

First published: November 2017

Production reference: 1021117

Published by Packt Publishing Ltd.
Livery Place
35 Livery Street
Birmingham
B3 2PB, UK.
ISBN 978-1-78328-387-3

www.packtpub.com

Credits

Authors
Shantanu Bhadoria
Ruben Oliva Ramos

Copy Editor
Stuti Srivastava

Reviewer
Soham Kamani

Project Coordinator
Virginia Dias

Commissioning Editor
Kartikey Pandey

Proofreader
Safis Editing

Acquisition Editor
Heramb Bhavsar

Indexer
Aishwarya Gangawane

Content Development Editor
Sweeny Dias

Graphics
Kirk D'Penha

Technical Editor
Vishal Kamal Mewada

Production Coordinator
Deepika Naik

About the Authors

Shantanu Bhadoria is an avid traveler and the author of several popular open source projects in Perl, Python, Golang, and Node.js, including many IoT projects. When in Singapore, he works on paging and building control systems for skyscrapers and large campuses in Singapore, Hong Kong, and Macau. He has authored and contributed to public projects dealing with control over gyroscopes, accelerometers, magnetometers, altimeters, PWM generators, and other sensors and controllers, as well as sensor fusion algorithms such as Kalman filters.

Shantanu's work in IoT and other fields can be accessed on his GitHub account with the name `shantanubhadoria`.

He is also the author of `Device::SMBus`, a popular Perl library used to control devices over the I2C bus.

> *I would like to thank the flying spaghetti monster (FSM) for showing me the guiding light of knowledge with his noodly appendage. I would like to thank my mother, sister, and my wife for supporting me while I was busy writing this book, and above all, I would like to express my affection for my dear pets (Dudette and Buddy) for cheering me up when I needed it! Ramen!*

Ruben Oliva Ramos is a computer systems engineer from Tecnologico of León Institute, with a master's degree in Computer and Electronic Systems Engineering, with a specialization in Teleinformatics and Networking from the University of Salle Bajio in Leon, Guanajuato, Mexico. He has more than five years of experience in developing web applications to control and monitor devices connected to Arduino and Raspberry Pi using web frameworks and cloud services to build the IoT applications.

Ruben is a Mechatronics teacher at the University of Salle Bajio and teaches students enrolled for the master's degree in Design and Engineering of mechatronics Systems. He also works at Centro de Bachillerato Tecnologico Industrial 225 in León, Guanajuato Mexico, and teaches subjects, such as electronics, robotics, and control, automation and microcontrollers, at mechatronics technician career. He is also a consultant and developer for projects in areas, including monitoring systems and datalogger data, using technologies, such as Android, iOS, Windows Phone, HTML5, PHP, CSS, Ajax, JavaScript, Angular, and ASP.NET; databases, such as SQLite, MongoDB, and MySQL; and web servers, such as Node.js and IIS. Ruben has done hardware programming on Arduino, Raspberry Pi, Ethernet Shield, GPS and GSM/GPRS, ESP8266, and control and monitoring systems for data acquisition and programming.

Ruben is the author of the following books, also by *Packt*:

- Internet of Things Programming with JavaScript
- Advanced Analytics with R and Tableau

I would like to thank my savior and lord, Jesus Christ for giving me strength and courage to pursue this project, to my dearest wife, Mayte, our two lovely sons, Ruben and Dario, To my dear father (Ruben), my dearest mom (Rosalia), my brother (Juan Tomas), and my sister (Rosalia) whom I love, for all their support while reviewing this book, for allowing me to pursue my dream and tolerating not being with them after my busy day job.

I'm very grateful with Packt Publishing for giving the opportunity to collaborate as an author and reviewer, to belong to this honest and professional team.

About the Reviewer

Soham Kamani is an author, developer, and an open source enthusiast. He has experience as a digital consultant and as a product engineer.

Soham actively contributes to open source projects and writes about topics ranging from hardware electronics to web development.

He has authored the book *Full Stack Web Development with Raspberry Pi*, with Packt.

www.PacktPub.com

For support files and downloads related to your book, please visit www.PacktPub.com.

Did you know that Packt offers eBook versions of every book published, with PDF and ePub files available? You can upgrade to the eBook version at www.PacktPub.com, and as a print book customer, you are entitled to a discount on the eBook copy. Get in touch with us at service@packtpub.com for more details.

At www.PacktPub.com, you can also read a collection of free technical articles, sign up for a range of free newsletters and receive exclusive discounts and offers on Packt books and eBooks.

https://www.packtpub.com/mapt

Get the most in-demand software skills with Mapt. Mapt gives you full access to all Packt books and video courses, as well as industry-leading tools to help you plan your personal development and advance your career.

Why subscribe?

- Fully searchable across every book published by Packt
- Copy and paste, print, and bookmark content
- On demand and accessible via a web browser

Customer Feedback

Thanks for purchasing this Packt book. At Packt, quality is at the heart of our editorial process. To help us improve, please leave us an honest review on this book's Amazon page at `https://www.amazon.com/dp/1783283874`.

If you'd like to join our team of regular reviewers, you can email us at `customerreviews@packtpub.com`. We award our regular reviewers with free eBooks and videos in exchange for their valuable feedback. Help us be relentless in improving our products!

Table of Contents

Preface

The Raspberry Pi was first introduced by the Raspberry Pi foundation to promote the education of basic computer science in developing countries. Little did the foundation know that they had truly started a DIY home project revolution. A development board running a flavor of the Debian Linux distribution at only $25 complete with a multitude of hardware pins providing support for GPIO, UART, I2C, SPI, and more, was a godsend for tinkerers hoping to wire it up to control sensors and actuators for their home automation ideas. With the introduction of version 3 and Raspberry Pi Zero W, programmers can connect to Wi-Fi with the Pi without the need to use a USB Wi-Fi dongle.

Arduino boards were first launched in the early 2000s. With a real-time operating system, and onboard ATMega controllers, these boards have always been easy to program, and they have provided a great way to interface with analog devices. Their ability to read analog signals without a 2D chipset and the extremely low power consumption of some variants allows these boards to complement the Raspberry Pi for automation projects. Just like the Raspberry Pi, these boards also provide the ability to communicate on 12C, SPI, and other protocols. Some specialized variants, such as the ESP8266, also support Wi-Fi connectivity.

As you will read through the chapters in this book you will notice how we use the Raspberry Pi or an appropriate variant of Arduino as per the specific needs of the chapter. We will show you how to use these boards to create real-life, cool, practical automation projects.

What this book covers

Chapter 1, *Creating a Raspberry Pi-Powered Magic Mirror*, teaches you about the Raspberry Pi development board, how to navigate the command line using basic Linux commands, and how to set up the open source Magic Mirror modular platform to work with the Pi. Once we set up, we will take a look at configuring the Magic Mirror and integrating third-party modules to create a customized smart mirror experience.

Chapter 2, *Automated Gardening System*, begins by explaining how to build a simple smart gardening system that automatically waters your plants as needed. The chapter explores the use of an always-on low power Arduino Pro Mini setup and outdoor waterproofing options. We will go into the specifics of building a system that senses when your plants need watering so that you can maintain ideal conditions for your garden to grow.

Chapter 3, *Integrating CheerLights into a Holiday Display*, reveals how to craft a festive dynamic light display using CheerLights, ESP8266, and NeoPixels. It also navigates you through the functionality of the ESP8266 breakout board and shows you how you can get the world to light up your festive display through Twitter tweets. These Wi-Fi-connected lights change colors based on tweets sent from around the world.

Chapter 4, *Erase Parking Headaches with OpenCV and Raspberry Pi*, uses OpenCV, a Raspberry Pi with the Wheezy distribution, the Amazon Web Service's Simple Notification System, and a webcam, to teach you how to create a notification system for a parking space.

Chapter 5, *Building Netflix's The Switch for the Living Room*, helps you design your own button, which, when pressed, will dim the lights, order pizza, turn on Netflix, and silence notifications on your phone—Netflix, being fans of the maker movement, put out a great project called The Switch. This is an excellent way to quickly get into the movie-watching mood, and it will be sure to impress any guests.

Chapter 6, *Lock Down with a Windows IoT Face Recognition Door System*, is for you, if you have ever wanted to create a locking system that relies on facial recognition. You will use ideas taken from Microsoft's Hack the Home initiative, the Raspberry Pi, an electric door strike, and other components to create a security system using Windows IoT Core for Raspberry Pi 3.

What you need for this book

For this book, the main component you will need is, of course, a control board, which, depending on the chapter, may be a Raspberry Pi or an Arduino-based board. The beginning of each chapter lists the complete list of items required to build your project. We will also show you how to program the control boards for the project and how to connect all the components together. We will guide you step-by-step through building the hardware so that you are not left behind.

An understanding of electronic hardware, networking, and basic programming skills would help you move quickly through the chapters, however, we will provide detailed explanations of the software setup in case you need more hand-holding at any point.

Who this book is for

This book is for you if you agree with one of these:

- You have seen cool projects on the internet that amateur beginners have set up in their homes and you have an interest in hopping on the DIY automation bus, but don't know where to start.
- You have used the Raspberry Pi for a media center or as a Linux desktop and you want to take it to the next level and find the answer to the question "what else can it do?"
- You want to impress your friends with your use of technology to simplify mundane everyday home tasks.
- You want someone to hold your hand through creating some cool home automation setups so that you may get the confidence to try your own big idea next.

All the projects in this book are examples of how you may use these boards. You may customize these ideas for your own situations; it depends on what you want to do or create. Your creativity and imagination is the limit.

Conventions

In this book, you will find a number of text styles that distinguish between different kinds of information. Here are some examples of these styles and an explanation of their meaning. Code words in the text, database table names, folder names, filenames, file extensions, pathnames, dummy URLs, user input, and Twitter handles are shown as follows: "Make sure your Raspberry Pi is updated and upgraded. You can do this by typing `sudo apt-get update && sudo apt-get upgrade`".

A block of code is set as follows:

```
{
  module: 'module name',
  position: 'position',
  header: 'optional header',
  config: {
    extra option: 'value'
  }
},
```

When we wish to draw your attention to a particular part of a code block, the relevant lines or items are set in bold:

```
{
  module: 'module name',
  position: 'position',
  header: 'optional header',
  config: {
    extra option: 'value'
  }
},
```

Any command-line input or output is written as follows:

```
wget http://node-arm.herokuapp.com/node_latest_armhf.deb
sudo dpkg -i node_latest_armhf.deb
```

New terms and **important words** are shown in bold.

 Warnings or important notes appear like this.

 Tips and tricks appear like this.

Reader feedback

Feedback from our readers is always welcome. Let us know what you think about this book-what you liked or disliked. Reader feedback is important to us as it helps us develop titles that you will really get the most out of. To send us general feedback, simply email feedback@packtpub.com, and mention the book's title in the subject of your message. If there is a topic that you have expertise in and you are interested in either writing or contributing to a book, see our author guide at www.packtpub.com/authors.

Customer support

Now that you are the proud owner of a Packt book, we have a number of things to help you to get the most from your purchase.

Errata

Although we have taken every care to ensure the accuracy of our content, mistakes do happen. If you find a mistake in one of our books—maybe a mistake in the text or the code—we would be grateful if you could report this to us. By doing so, you can save other readers from frustration and help us improve subsequent versions of this book. If you find any errata, please report them by visiting http://www.packtpub.com/submit-errata, selecting your book, clicking on the Errata Submission Form link, and entering the details of your errata. Once your errata are verified, your submission will be accepted and the errata will be uploaded to our website or added to any list of existing errata under the Errata section of that title. To view the previously submitted errata, go to https://www.packtpub.com/books/content/support, and enter the name of the book in the search field. The required information will appear under the Errata section.

Piracy

Piracy of copyrighted material on the internet is an ongoing problem across all media. At Packt, we take the protection of our copyright and licenses very seriously. If you come across any illegal copies of our works in any form on the internet, please provide us with the location address or website name immediately so that we can pursue a remedy. Please contact us at copyright@packtpub.com with a link to the suspected pirated material. We appreciate your help in protecting our authors and our ability to bring you valuable content.

Questions

If you have a problem with any aspect of this book, you can contact us at questions@packtpub.com, and we will do our best to address the problem.

1
Creating a Raspberry Pi-Powered Magic Mirror

We'll be operating with the understanding that this may be the first Raspberry Pi project that you have attempted to undertake. The **Magic Mirror** is a practical and easy way to introduce yourself to working the Raspberry Pi and will also serve a great conversation piece for your home. To briefly cover the contents of this chapter, you will learn about the Raspberry Pi single-board computer and how to navigate the command line using basic Linux commands. We will focus on downloading the latest version of the Magic Mirror project by Michael Teeuw from GitHub (`https://github.com/MichMich/MagicMirror`) and setting up the open source modular platform to work with the Pi 3. Once downloaded, we will take a look at editing the content for the Magic Mirror and how one might go about integrating third-party modules to create a custom Magic Mirror experience. As a final step, a discussion will surround the construction of the mirror's frame and what might be best for your personal home experience.

By the end of this chapter, you'll know how to:

- Work with the Raspberry Pi
- Operate within LXTerminal
- Navigate the Magic Mirror repository
- Do basic file editing with GNU Nano
- Use the Raspberry Pi GPIO to attach sensors

What is the Raspberry Pi?

The Raspberry Pi is a credit-card-sized single-board computer that was developed by the Raspberry Pi Foundation in 2012. The foundation's main goal is to promote computer literacy across the globe by offering an affordable and mutable bit of hardware to the masses. It's become a huge hit among the maker communities and is paving its way through education as a cheap, practical, and convenient way to teach digital and physical making. As we move through the text, I will be centering my conversation on the base model and necessities of a Raspberry Pi 3, as this is the most current version of the Raspberry Pi on the market. If you're using a Raspberry Pi 2 or B+, I'll make sure to point out the differences in hardware usage as we move along.

While taking a closer look, the Raspberry Pi 3 boasts of some impressive specs for its size:

- A 1.2 GHz 64-bit quad-core ARMv8 CPU
- 802.11n Wireless LAN
- Bluetooth 4.1
- **Bluetooth Low Energy (BLE)**
- 1 GB RAM
- Four USB ports
- 40 GPIO pins
- Full HDMI port
- Ethernet port
- Combined 3.5mm audio jack and composite video
- **Camera Interface (CSI)**
- **Display Interface (DSI)**
- Micro SD card slot
- VideoCore IV 3D graphics core

Furthermore, the Raspberry Pi 3 has the same form factor as the earlier models, the Raspberry Pi 2 and the B+, allowing you to reuse the casing and accessories.

Raspberry Pi Model 3:

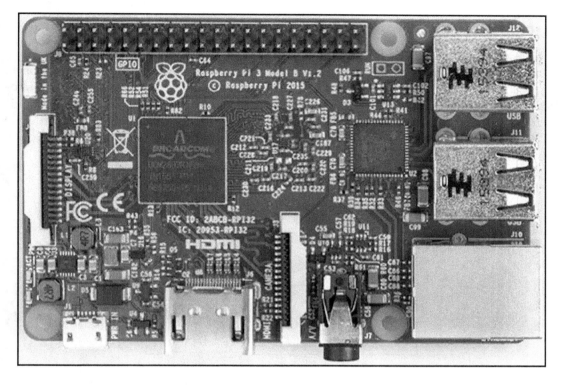

You will need the following materials for this project:

- **Project materials**:
    - Raspberry Pi 2/3
    - Micro USB charging cable
    - Wi-Fi dongle (if you are using Pi 2)
    - A microSD card
    - Monitor (HDMI/VGA)
    - Two-way glass/acrylic
    - Wooden frame
- **Optional materials**:
    - PIR motion sensor
    - Pi Camera Module
    - Ultrasonic sensor—HC-SR04
    - HDMI-to-VGA converter

Purchasing the Raspberry Pi

When purchasing a Raspberry Pi, it may look a bit intimidating at first, given the sheer amount of add-ons or the variety of kits from which you can choose. At a minimum, for this project and the others in this book, you'll need a kit that contains a Raspberry Pi 3, a micro-USB charging cable, an 8 gigabyte (or larger) microSD card preinstalled with NOOBS, and an HDMI cable. For the novice user, it's generally recommended that you also acquire an HDMI monitor and a USB-connected keyboard and mouse to easily display and interact with the Raspberry Pi interface:

The Raspberry Pi revision 2 and B+ models differ from the Raspberry Pi revision 3, as neither revision 2 or B+ have Bluetooth or Wi-Fi built into the board itself. So, if you're working with either of these models, be aware that you will have to purchase a Wi-Fi dongle (highly recommended) or a Bluetooth dongle in order to have these connectivity options available to you.

Setting up the Raspberry Pi

Assuming that you've gotten the required materials, it's now time to get the Raspberry Pi up and running. Before powering it on, you're going to make sure that you connect the Pi to a monitor, connect the keyboard and mouse, and make sure that the microSD card is inserted. If you have any dongles to plug into the extra USB ports, now would be the time to do that.

In discussing the microSD card, with most kits bought online for the Raspberry Pi, the microSD card will come preloaded with **NOOBS** (known as **New Out Of the Box Software**), which lets you easily set up Raspbian, the Foundation's officially supported operating system. Raspbian is a Linux distribution based on the Debian distribution. You can, however, purchase a microSD card that does not have the NOOBS installation preloaded on it. If this is the case, using your home computer or laptop, you can manually install NOOBS or the image of your choosing onto the SD card. Here are the steps for the manual installation, as aligned to the Raspberry Pi Foundation's suggestions:

1. Visit the SD Association's website (`https://www.sdcard.org/`) and download SD Formatter 5.0 for either Windows or Mac systems.
2. Follow the instructions to install the software.
3. Insert your SD card into the computer or laptop's SD card reader and make note of the drive letter allocated to it, for example, F or G.
4. In **SD Formatter**, select the drive letter for your SD card and format it.
5. Download the NOOBS ZIP file from the Raspberry Pi Foundation Downloads page (`https://www.raspberrypi.org/downloads/noobs/`).
6. Save this to a folder on your computer and then extract the files.
7. Once your SD card has been formatted and the files from the ZIP are extracted, you are going to drag the unzipped files onto the SD card. As a note, make sure you're dragging the contents of the NOOBS folder onto the SD card. If you move the entire NOOBS folder itself, the installation process will not follow through.
8. When the files have finished transferring, eject the SD card and place it into your Raspberry Pi.

Assuming the microSD card has made it to your Raspberry Pi and all of the requisite hardware is connected, it is now time to power on the board by plugging in the micro USB charging cable.

This will take a little time as you go through the setup screen, but you'll be looking at the Raspbian desktop shortly:

Moving on to the LXTerminal

While the Raspberry Pi has a healthy variety of applications and programs for our consumption, we're going to be primarily using the LXTerminal to download and interact with the Magic Mirror program. For those of you unfamiliar with the Terminal, it's essentially a program that allows a user to directly manipulate their computer system through the use of commands. As we move through the text, when I refer to the *command line*, I'm talking about the Terminal and the location where a user will input the commands.

Understanding how the Terminal works and being able to use the command line is a very powerful tool when working to manipulate files on the Raspberry Pi. Commands can be strung together and input in order to efficiently carry out tasks that other applications cannot. When you initially open LXTerminal, you will see a screen with a blinking cursor. On the left-hand side, there will be a prompt that shows your current username and the hostname of the Pi. Both the username and hostname can be changed later on in settings if you'd like to tailor the Pi further for your uses:

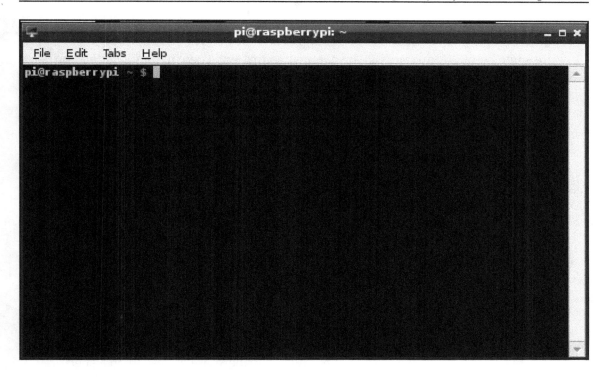

Basic Linux commands

There are basic Linux commands that are going to be very useful in your navigation through your filesystems through the Terminal. In this section, I want to list some essential commands that will help you when building the Magic Mirror. With each command, I'll provide a brief explanation. As with anything, the best way to learn how to successfully use these commands is with practice. Also, note that syntax is important when working with the command line. The computer interprets capitalized and uncapitalized letters as two different symbols, so we have to make sure we type our commands as they are shown. A good rule of thumb is that if something doesn't work the first time, go back and proofread what you have typed.

Helpful commands

Here are some helpful Linux or NOOBS commands that will come in handy when you work on your Raspberry Pi:

- sudo: Short for **super user do**, this command is powerful and necessary. When used before any others in the command line, you're telling the computer that you are running the command as the root user. This gives you the ability to alter files that may be unalterable to the regular user on the computer. When first working with the Terminal, I suggest you use sudo primarily, as opposed to su (though su is needed for particular actions). This will curtail your use of the super user command, putting a slight speed bump between you and any commands that may unintentionally delete necessary files from your system.

- su: While similar to sudo, instead of running a command as root, su makes you root. As I said earlier, sudo is useful for targeted commands, but if you're working extensively as root, su may be the better option. Just make sure you stay aware that you're working as a root user, as you do not want to delete anything necessary in your Raspberry Pi system files.

- cd: This changes your location to a directory of your choosing. Remember, a directory is essentially a folder, and this allows you to navigate to a specific folder as long as you know the pathway. Furthermore, if you are in a specific directory and want to return to your /home directory, typing in cd will bring you immediately back to /home.

- cd –: To build on the cd command, a useful command is to add the *dash* after the *space* after cd. This will allow you to immediately return to the previous directory in your pathway, for example, cd /MagicMirror/modules/. This will bring you to the modules folder within the MagicMirror folder. If you want to go back to the MagicMirror folder, you would then type cd –.

- ls: When you're in a directory and wish to know the folder's contents, typing in this command will list the files within the folder.

- pwd: This stands for **print working directory**. It is very useful if you've forgotten your current pathway or need to access that information.

- mkdir: This command enables you to create a directory with a name of your choosing.

- cp: This is the copy function. It allows you to copy a file from its source destination to another destination of your choosing.

- nano: This command calls upon the Terminal's nano text editor, enabling you to access and modify the contents of the file from within the Terminal emulator. This is very useful for quick changes to files on the Pi, and I will be using this as the command-line editor for this text.

Let's move forward. I find these to be the most useful commands for this project; however, there is a vast array of commands that will make your navigation more efficient. As you learn more with the Pi, it's encouraged that you research more commands and how they work in order to maximize your experience in LXTerminal.

The Magic Mirror

The project was first put together by Michael Teeuw in 2014 and has since garnered much popularity in the maker community, inspiring smart mirror projects created by the average laymen. The project is open source and community-driven, enabling all users to download, edit, and create using this framework. On the most basic level, the Magic Mirror program is run on the Raspberry Pi, typically connected to a monitor specifically designated for this project.

The user then programs the Magic Mirror interface to his or her liking, adding functions such as e-mail notifications, NFL scores, weather alerts, reminders, and additional text. Once the Magic Mirror is set up to your liking, it's common to see people install a wooden frame and a two-way mirror around the monitor, effectively creating a do-it-yourself smart mirror.

Downloading the Magic Mirror repository

Now that we've established how to set up our Raspberry Pi and familiarized ourselves with the Terminal, it's time to download the Magic Mirror repository. In order to do this, you'll have to clone the repository that is hosted on the popular website GitHub. GitHub is an online version control repository, basically, a place where the changes to computer files and code can be tracked and all workers on a particular project can come together to add the appropriate documentation. In line with the open source nature of this project, it's very helpful for your future endeavors with this project. If you find that you have further questions, bug issues, or code to contribute, checking out their GitHub repository is the next and later step.

Now in terms of cloning the repository, using this phrasing is deliberate. To clone a repository means to not only download the current working copy of a project, but to also download every version of the project available. This is extremely useful for individuals who face problems with file corruption or server issues. So, moving into cloning the Magic Mirror repository, it's fairly straightforward and we'll get to it shortly. I'm having you walk through the manual installation of the Magic Mirror so that you have a better understanding of what is happening while interacting with the command line. I'll show the alternate way later on in the text, but this builds some necessary skills for later on in this book. In turn, we have to install a few more packages before cloning into the repository in order to have the Magic Mirror run efficiently.

Installing Node.js

Node.js is an important aspect of the Magic Mirror setup as it creates the screen image with the specific modules on the monitor. It's described as *"an asynchronous event driven JavaScript runtime... designed to build scalable network applications"* (about Node.js). While Raspbian, our operating system, does a timely job at staying updated, it behooves us to make sure that we have the latest version of Node.js in order to run our Magic Mirror program. We're going to install the latest 64-bit ARM (known as **A**dvanced **R**educed Instruction Set Computer **M**achines) version for our Raspberry Pi. In order to install the latest version of Node.js, you're going to execute the following steps:

1. Make sure that you have an active Internet connection.
2. Open up LXTerminal.
3. In the Terminal, type the following:

```
wget http://node-arm.herokuapp.com/node_latest_armhf.deb
sudo dpkg -i node_latest_armhf.deb
```

4. This was me telling the Pi to grab the most current version of Node.js and then switch to the root user in order to install the node package using dpkg, which is the Debian package management system.
5. Wait for Node.js to download.
6. Once the download has finished, double-check that you have the latest version by inputting node -v into the command line.

Installing Grunt

With some of the updates to the Magic Mirror program, I've found that it's best to have the latest version of Grunt installed on your Pi as well. Grunt is basically a task manager and a build manager that integrates with Node.js. It has certain features that are helpful in the JavaScript environment, such as minification or the process of removing all unnecessary characters from source code without changing its functionality. On top of that, Grunt enables your program to reload as you make changes to your code during development. You'll notice, when we start the Magic Mirror later on, that we're going to call on Grunt usage with the `npm start` command. Grunt installation is very simple. We're going to want to install Grunt's command-line interface globally on the Pi. This way, we can readily call upon it when working from any directory:

```
sudo npm install -g grunt-cli
```

With Node.js and Grunt now downloaded, it's time to finally clone into our repository:

1. Make sure your Raspberry Pi is updated and upgraded. You can do this by typing `sudo apt-get update && sudo apt-get upgrade`.
2. Clone the repository with `git clone https://github.com/MichMich/MagicMirror ~/MagicMirror`.
3. This will download the directory containing the Magic Mirror files directly to your `/home/pi` folder.
4. You're going to enter the repository with `cd ~/MagicMirror`.
5. You are then going to install and run the application using `npm install && npm start`.

It will take a second or two, but you will then see the default screen for your Magic Mirror pop up on your screen (Teeuw).

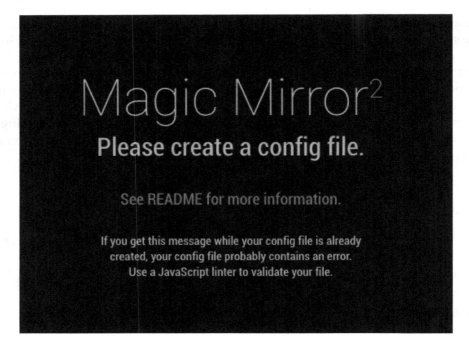

Working with the configuration file

Now that we have the Magic Mirror up and running, we have to personalize it. Consider where the Magic Mirror is going to be placed within your home. Are you going to be using it as a clock? Does it replace your bathroom mirror? Is this something that you've mounted near the front door so that you may scan some information as you leave for the day? All are excellent questions to consider and will mold how you choose which modules to include and which information should be highlighted. As you consider your options, take note of the default modules that are preinstalled with the application:

- **Clock**: This displays the current date and time.
- **Calendar**: This displays information from any public iCal calendar and can combine multiple calendars into one.

- **Current weather**: This will display your chosen location's weather information. It's geared toward using the API from http://www.openweathermap.org/api.
- **Weather Forecast**: Also using OpenWeatherMap, this will display the location's forecast for the week.
- **News Feed**: You can input news URLs of your choosing and the title of each news source, and this will scroll through the major headlines of the chosen news sources.
- **Compliments**: This gives you the ability to randomize your own text content on the screen at a given the time of day: morning, afternoon, and evening.
- **Hello world**: This places static text on your screen in a location of your choosing.
- **Alert**: This module will display notifications from other modules.

As you can see, these working in tandem can create a very informative mirror for your everyday use. So, how do we actually customize? The customization process relies on the configuration file nestled in /MagicMirror/config. You'll notice that when you move into the config directory, a file that you see listed is config.js.sample:

```
shantanus-MacBook-Air:MagicMirror shantanubhadoria$ ls config/
config.js.sample
```

We have to make a copy of config.js.sample, and the easiest way to do that will be by making sure that you're in the config directory and making a copy of the file:

```
cp config.js.sample config.js
```

So, any changes that you make to the config.js file will show up on the Magic Mirror screen. If you make an inexcusable error or find yourself needing to revert to the defaults entirely, you still have the sample copy to rely upon.

This is the sort of screen you will see with the default config.

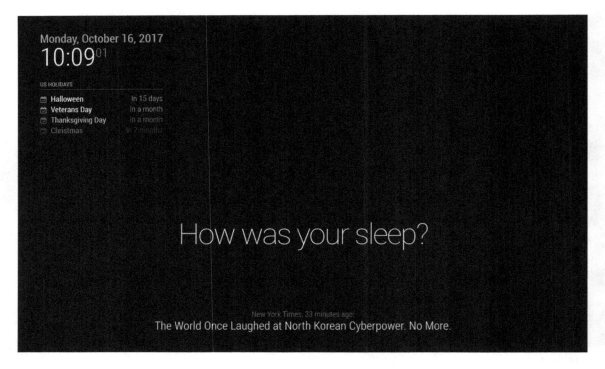

Editing the configuration file

If you're a complete novice to JavaScript, don't worry too much about this process. Remember, the idea is to learn through trial and error and create a product that instills some pride. One way to build skills is to start editing and adding to the configuration file copy that you created in the previous section. The nice part about editing this file is that if your syntax is wrong or there's a typo living somewhere in your document, the Magic Mirror will display a screen that alerts you to this problem. So there's no harm in trying it out!

If you find it difficult to fix any problems or errors in the script, be mindful that there are browser-based JavaScript editors online that help you find the errors in your code. It comes down to a matter of copying and pasting, so I'd encourage you to turn to a website such as JSLint, found at `http://www.jslint.com/`. So, on to the configuration file as a whole:

```
/* Magic Mirror Config Sample
 *
 * By Michael Teeuw http://michaelteeuw.nl
 * MIT Licensed.
 *
 * For more information how you can configurate this file
 * See https://github.com/MichMich/MagicMirror#configuration
 *
 */

var config = {
    address: "localhost", // Address to listen on, can be:
                          // - "localhost", "127.0.0.1", "::1" to listen on
```

```
loopback interface
                        // - another specific IPv4/6 to listen on a
specific interface
                        // - "", "0.0.0.0", "::" to listen on any
interface
                        // Default, when address config is left out, is
"localhost"
    port: 8080,
    ipWhitelist: ["127.0.0.1", "::ffff:127.0.0.1", "::1"], // Set [] to
allow all IP addresses
                                                // or add a
specific IPv4 of 192.168.1.5 :
                                                // ["127.0.0.1",
"::ffff:127.0.0.1", "::1", "::ffff:192.168.1.5"],
                                                // or IPv4 range
of 192.168.3.0 --> 192.168.3.15 use CIDR format :
                                                // ["127.0.0.1",
"::ffff:127.0.0.1", "::1", "::ffff:192.168.3.0/28"],

    language: "en",
    timeFormat: 24,
    units: "metric",

    modules: [
        {
            module: "alert",
        },
        {
            module: "updatenotification",
            position: "top_bar"
        },
        {
            module: "clock",
            position: "top_left"
        },
        {
            module: "calendar",
            header: "US Holidays",
            position: "top_left",
            config: {
                calendars: [
                    {
                        symbol: "calendar-check-o ",
                        url:
"webcal://www.calendarlabs.com/templates/ical/US-Holidays.ics"
                    }
                ]
            }
```

```
        },
        {
            module: "compliments",
            position: "lower_third"
        },
        {
            module: "currentweather",
            position: "top_right",
            config: {
                location: "New York",
                locationID: "", //ID from
http://www.openweathermap.org/help/city_list.txt
                appid: "YOUR_OPENWEATHER_API_KEY"
            }
        },
        {
            module: "weatherforecast",
            position: "top_right",
            header: "Weather Forecast",
            config: {
                location: "New York",
                locationID: "5128581", //ID from
http://www.openweathermap.org/help/city_list.txt
                appid: "YOUR_OPENWEATHER_API_KEY"
            }
        },
        {
            module: "newsfeed",
            position: "bottom_bar",
            config: {
                feeds: [
                    {
                        title: "New York Times",
                        url:
"http://www.nytimes.com/services/xml/rss/nyt/HomePage.xml"
                    }
                ],
                showSourceTitle: true,
                showPublishDate: true
            }
        },
    ]

};

/*************** DO NOT EDIT THE LINE BELOW ***************/
```

While reprinted here, you can also access this file for editing by employing the use of the nano function in the command line. As always, make sure that you're in the configuration directory found within the Magic Mirror directory: `/home/pi/MagicMirror/configuration`. At this point, you want to tell the Pi that you wish to edit `config.js`. In order to do that, type `sudo nano config.js` into the command line. This will bring up the preceding file text on your screen, where you can easily move the cursor around to add and remove items.

In terms of observation, there are a few items to which I wish to call your attention. Toward the top of the script, you'll notice that there are some initial customizable options, such as `port`, `ipWhitelist`, `language`, `timeFormat`, and `units`. Let's explore each briefly:

- `port`: Not to be confused with a USB port, this is regarded as the endpoint of information in an operating system. You can change the port address, but `8080` is the conventional address that you'll see used somewhat ubiquitously as your personally hosted web server. Over your local network, it's easily accessed and can be used to view your Magic Mirror in a browser, as opposed to the application view. As a side note, running the Magic Mirror on a local host would be useful if you wanted multiple browser displays showing the same information. Some museums and venues use this setup for guests.
- `ipWhitelist`: If you'd like to remote into your Magic Mirror, you'll have to add the IP address of the machine you'd like to use. With updates to the Magic Mirror happening frequently, I would suggest that you look to the active Magic Mirror community and its discussions in case there are any issues.
- `language`: While setting to English, you also have the options of nl, ru, and fr, Dutch, Russian, and French, respectively.
- `timeFormat`: You can change this from the 24-hour clock format to the 12-hour format by inputting `12` in this space.
- `units`: While setting to metric, you can replace this with imperial.

After going through these preferential settings, you'll notice that the body of the script deals directly with the modules that are going to be displayed on your screen. You'll notice a pattern with the curly brackets, as they enclose each module. There is a standardization to how each module is set up, which may be helpful if you find yourself puzzled about how to incorporate preferences and variables:

```
{
  module: 'module name',
  position: 'position',
  header: 'optional header',
  config: {
```

```
      extra option: 'value'
  }
},
```

Understanding the module

So, let's take the `compliments` module as our example working piece. If you pass your eyes back over the `config.js` file, you'll notice that the `compliments` module is nestled fifth from the top. You'll notice that the only information in the `config.js` file for the `compliments` module is its placement: lower third. How does the `compliments` module know what information to display? Where is the revolving text coming from? The `config.js` file is using and calling up the information stored in the `compliments` directory, found in `/home/pi/MagicMirror/defaults/compliments`, specifically, `compliments.js`. Logically, we would need to edit the `compliments.js` file in order to change the content of the `compliments` module. Accordingly, we must understand that the `config.js` file literally deals with the spacing and *configuration* of the Magic Mirror display. Moving forward, let's take a look at a portion of the `compliments.js` file after you've navigated to the `compliments` directory:

```
cd /home/pi/MagicMirror/defaults/compliments
sudo nano compliments.js
```

While looking at this file, we see the most important content aspects—anytime, morning, afternoon, and evening, toward the top. As you can guess, these compliments or text portions roll through the screen given the appropriate times of the day. Consider what content you would like to change or add and play around with modifying it. Also, make note that later in the `compliments` file, you have the ability to modify the hours dedicated to morning and afternoon, the leftover hours being dedicated to the evening:

```
complimentArray: function() {
    var hour = moment().hour();
    var compliments;

    if (hour >= 3 && hour < 12 &&
this.config.compliments.hasOwnProperty("morning")) {
        compliments = this.config.compliments.morning.slice(0);
    } else if (hour >= 12 && hour < 17 &&
this.config.compliments.hasOwnProperty("afternoon")) {
        compliments = this.config.compliments.afternoon.slice(0);
    } else if(this.config.compliments.hasOwnProperty("evening")) {
        compliments = this.config.compliments.evening.slice(0);
    }
    if (typeof compliments === "undefined") {
```

```
        compliments = new Array();
    }
    if (this.currentWeatherType in this.config.compliments) {
        compliments.push.apply(compliments,
this.config.compliments[this.currentWeatherType]);
    }
    compliments.push.apply(compliments, this.config.compliments.anytime);
    return compliments;
},
```

Once you've played around with modifying, understand that this process is replicated for the majority of modules in this program. I'd recommend that you take the time to look through the default folders and see what else you can change. If you're worried about mistakes or editing a file in an error, make a copy of the original file, rename it, and leave it so that you have a reference point or a file to revert to.

Installing third-party modules

While the Magic Mirror comes with default modules, there are industrious Internet folk out there creating new, open source modules for download and addition by going to the third-party module portion of the Magic Mirror GitHub website, `https://github.com/MichMich/MagicMirror/wiki/MagicMirror%C2%B2-Modules#3rd-party-modules`. While extensive to list, you'll note that there are modules for Twitter, Instagram, NFL scores, and Amazon's Alexa, to name a few. These come with varying degrees of skill needed to set up, so it's suggested that you assess which seems to be the best to try and then work through the installation and setup steps. Importantly, when installing a third-party module, you're going to clone the repository into the Magic Mirror modules folder. Your steps would involve the following:

1. Navigate to the Magic Mirror modules folder.
2. Determine the URL of the GitHub repository you wish to clone.
3. Clone the repository using the steps mentioned previously in the chapter.

Here is an example of what your mirror may look like once you place a two way mirror on the screen plugged into your Raspberry Pi 3:

Summary

In this chapter you learned how to set up a cool Magic Mirror that updates you on the news, shows you your local weather and so on while you check yourself out in the mirror. You also learned the basics of Raspberry Pi and the NOOBS operating system. You can enhance your mirror using motion sensors to activate your mirror when you are nearby. You can now continue to explore the Magic Mirror repository for some exciting modules to explore how others are using their Magic Mirror.

2
Automated Gardening System

We all love having a nice garden in our house, whether it's a tiny herb garden in a pot, a kitchen garden in your backyard, or a bed of flowers to welcome guests at the entrance. However, plants are living things and they require constant care and if you forget about them long enough they wither away. If you are like me in any way, you might have the same problem of killing your plants when you missed watering them due to forgetfulness or when you left home for a vacation.

In this chapter, we will create an automated gardening system that keeps an eye on your plant health by monitoring the soil humidity levels and watering your plants whenever the soil humidity level goes too low. This system will avoid the pitfall of a schedule-based gardening system which might end up killing your plants by over-watering them on rainy days.

We will discuss the following topics:

- Items required for this project
- Setting up a gardening system code—reading humidity
- Connecting it all together

Items required for the project

The gardening system has three parts: the control system, the humidity sensor that measures the humidity content of the soil, and a separate irrigation mechanism that irrigates the plants as required. We will use low-power components for our system (other than the irrigation pump) so that we can run them 24/7 without worrying about energy costs and they will also give us the flexibility of powering our system with batteries.

Waterproof junction box

You probably have a garden outdoors. You might want to keep your outdoor electronics near your garden as well; it will help to have a junction box that protects your electronics from rain and similar acts of nature. You can use hot-melt glue or quick-setting silicone paste to fix the electronics to the junction box as well as for providing some waterproofing:

Arduino Pro Mini 5v ATMEGA328P

We decide to go with the Arduino Pro Mini as our control board as we have decided to make our gardening system fully self contained and we don't need any external input to request for watering our plants, so we do not need a more expensive Arduino board with a Wi-Fi module. The Arduino Pro Mini is one of the cheapest Arduino boards with one of the lowest power consumptions around and at 5V supply voltage allows us the convenience of using the same source to power your Arduino and the relays.

The Arduino Pro Mini has an amperage of 23μA (0.023 mA) as compared to Arduino UNO, which uses around 45mA. This will give you awesome battery life if your setup is battery powered. While there are many fakes on the market that seem to work fine, it's recommended that your purchase yours from the Arduino store to support the Arduino project. Be sure to purchase the 5V version and not the 3.3V version (`https://store.arduino.cc/usa/arduino-pro-mini`):

USB FTDI connector

You will need a USB FTDI connector to connect the Arduino Pro Mini to your computer. This is required for you to program your Arduino Pro Mini. One option to purchase this cable is at Adafruit. Fortunately, the pins are mostly aligned (`https://www.adafruit.com/product/70`):

5V power supply

You will need a 5V power supply to power your setup. You can use an old cell phone charger as all cell phones with USB charging are 5V rated.

If you prefer to use battery power, a neat trick is to power your setup with a USB power bank like the ones used to charge phones. USB has a voltage rating of 5V and an amperage of more than 1V, which is more than enough to power our setup. When the battery goes down, you can simply charge your power bank again and continue using it. Adafruit sells a small power bank, but you can find one of these at any neighborhood cell phone stores (https://www.adafruit.com/product/1959):

In both of the previous cases, you will also need a USB cable to connect to the power bank or cell phone charger. You will need to strip the USB cable at the other end and find the VCC and GND wires. You can do this using a multimeter. In general, if you see a red wire, it's usually the 5V VCC and the black wire is usually the GND.

5V relay

Relays are electronically operated switches. Many relays use an electromagnet to mechanically operate a switch, but other operating principles are also used, such as solid-state relays. The relays are generally used to control or switch on/off a high-power circuit using a separate low-power control signal. That is, in the simplest form, a constant voltage applied across the input pins of a relay will toggle it, that is, either close (**connect**) or open (**disconnect**) another circuit. The relays which close (connect) a circuit on application of voltage are called **normally open relays** and the relays which open (connect) a circuit on application of voltage are called **normally closed relays**. Most modern relay circuits have both a normally open and normally closed connection.

For our purposes, we want our irrigation system to run only when we detect low soil humidity and be off otherwise, so we will connect our irrigation system across the normally open connection of the relay.

We will be using a relay with 5V control circuit voltage and with a switched circuit voltage rating of up to 220V:

Soil humidity sensor (hygrometer)

A soil humidity sensor essentially uses electrodes to measure the conductivity of the soil to calculate the humidity: the higher the soil humidity, the higher the conductivity. There are many soil humidity sensors available on the market. Since Arduino's analog pins, we can actually get the conductivity of the soil and take a decision based on the conductivity measurement of the sensor:

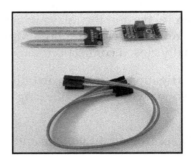

Photoresistor

A photoresistor is a light-activated variable resistor. The resistance of a photoresistor increases with decreasing light intensity and vice versa. We use a photoresistor to detect sunlight and avoid watering our plants at night:

Submersible pump

A submersible pump allows you to pump water from a water reservoir (or a bucket) at low level. Depending on where your reservoir is located, select a pump of appropriate power. An aquarium pump of sufficient power can be used if the reservoir is at the same or higher level as the garden. For small needs, a 40 watt pump would be sufficient. We will drop our submersible pump into the water reservoir and connect one power cord across the 5V relay to the power socket. This way, we will switch the pump on using our control circuit connected to the 5V relay inputs:

Drip irrigation system

A drip irrigation system allows you to slowly irrigate your garden while specifically targeting the roots of your plants and conserving water. It gradually increases the humidity of the soil drip by drip allowing your sensors enough time to trigger at the right moment and cut off the irrigation at the right moment once the soil achieves desired humidity:

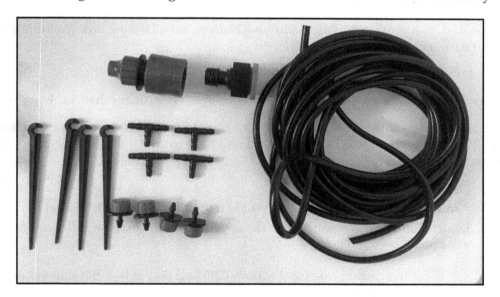

Koram IR-D blank distribution tubing watering drip kit on Amazon:
https://www.amazon.com/IR-D-Distribution-Tubing-Watering-4-Inches/dp/B013JPIJG4/.

Water reservoir

You must have a large enough reservoir to store enough water and for keeping your submersible pump in.

Setting up gardening system code (reading humidity)

In this section, we will set up the Arduino IDE and read humidity from the humidity sensor and print it at the Arduino serial port.

Setting up the Arduino IDE

The Arduino IDE is a free IDE provided by https://www.arduino.cc/ that makes it easy to write and deploy code on Arduino and Arduino-based boards like the Arduino Pro Mini. Before we start, you must install the Arduino IDE from the following link. You might have to select the correct version for your OS:

https://www.arduino.cc/en/main/software

If you already have the Arduino IDE, make sure its version is higher than 1.6.4.

You will have to switch the Arduino IDE to your board type. These instructions are for the Arduino Pro Mini with ATMEGA 328 and you will have to adjust these depending on the board you use; to do so, follow these steps:

1. Open up your Arduino IDE.
2. Arduino Pro or Pro Mini will be visible under **Tools** | **board:***. Select it.
3. Under **Tools** | **processor:*** select the appropriate processor and frequency for your board.

 These instructions are for Arduino IDE 1.8.1 and may differ depending on your version of IDE.

Once this setup is done, connect the Arduino Pro Mini to the USB port on your computer using a USB-to-FTDI cable.

Testing the Arduino Pro Mini

You can test the Arduino Pro Mini by uploading the standard LED blink code. In the Arduino IDE, click on **New** and paste the following default Arduino blink code in it:

```
// The setup function runs once when you press reset or power the board
void setup() {
    // initialize digital pin LED_BUILTIN as an output.
    pinMode(LED_BUILTIN, OUTPUT);
}

// The loop function runs over and over again forever
void loop() {
    digitalWrite(LED_BUILTIN, HIGH);    // turn the LED on
    delay(1000);                        // wait for a second
    digitalWrite(LED_BUILTIN, LOW);     // turn the LED off
    delay(1000);                        // wait for a second
}
```

Connecting the Arduino Pro Mini to the USB FTDI connector (FT232RL) and your computer

Connect the Arduino Pro Mini to the USB FTDI connector(FT232RL):

1. Connect your GND on the Arduino Pro Mini to the GND on your USB FTDI connector.
2. Connect the V+ on the Arduino Pro Mini to V+ or VCC on your USB FTDI connector.
3. Connect Tx on the Arduino Pro Mini to Rx on your USB FTDI connector.
4. Connect Rx on the Arduino Pro Mini to Tx on your USB FTDI connector.

The USB FTDI connector pins actually line up against Arduino Pro Mini pins so you may directly connect it to the Arduino Pro Mini if you have female headers on the FTDI connector and male headers on the Arduino Pro Mini or vice versa.

Now connect the USB FTDI connector (FT232RL) to a USB port on your computer:

Deploying your code

Deploy the aforementioned sketch on your Arduino Pro Mini using the deploy button, which is usually an arrow on the new versions of the Arduino IDE.

If the LED light on the Arduino Pro Mini blinks with a 1-second period between blinks, that means your board is working just fine.

Reading the humidity sensor

Now it's time to connect a humidity sensor to the Arduino and read in humidity values from it. Keep the USB FTDI connector connected to the Arduino Pro Mini and to the computer for now.

Connect the humidity sensor to the Arduino Pro Mini

The humidity sensor has at least three pins: the VCC pin, the GND pin, and the analog output pin usually labelled **AO**. There might be an additional pin labelled **DO** for digital output, which we will not use. Follow these steps:

1. Connect the VCC pin of the humidity sensor to the spare VCC pin on the Arduino Pro Mini.
2. Connect the GND pin of the humidity sensor to the spare GND pin on the Arduino Pro Mini.

3. Connect the AO pin of the humidity sensor to the analog input 0 pin on the Arduino Pro Mini:

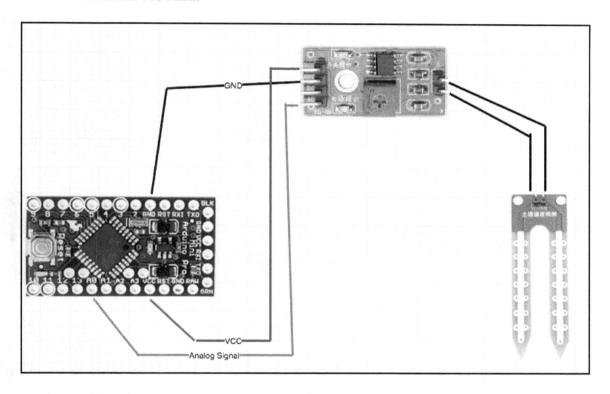

Uploading the sketch and testing humidity measurements

At this point, we have connected the humidity sensor to the Arduino. Keep a pot of soil handy to test our humidity measuring system:

```
/* Gardening system
 * Lets water our plants on time
 */

// humidity sensor is at analog port 0
const in SENSOR_PORT = A0;

int sensorValue = 0;

void setup() {
```

```
  // We will initialize the serial port so we can print humidity data to
the serial port.
  Serial.begin(9600);
}

// the loop routine runs over and over again forever:
void loop() {
  // Read the input on analog pin 0:
  sensorValue = analogRead(SENSOR_PORT);
  // A well watered soil will give a reading of atleast 380 to 400, a fully
moist soil will have reading of 1023.
  sensorValue = constrain(sensorValue, 400, 1023);
  // We will reverse the range so that we get a rough humidity percentage
reading i.e.
  // fully moist soil will give a reading of 100% for 100% humidity and
  // dry soil will read 0% for 0% humidity
  sensorValue = map(sensorValue,400,1023,100,0);

  // Print humidity value
  Serial.println(sensorValue);
  if (sensorValue >= 20) {
    Serial.println("soil is wet");
  } else
  {
    Serial.println("soil is dry");
  }

  delay(1000);
}
```

Upload the sketch to Arduino and start reading the serial port using the Arduino IDE by pressing .

Start placing the soil sensor in soil samples with different moisture levels. You should see the different humidity levels on the serial port monitor between 0 to 100.

Next we will set up the rest of our gardening system.

Connecting the pump power and control

Connect the submersible pump to a 220V power source via the normally open switch on the relay:

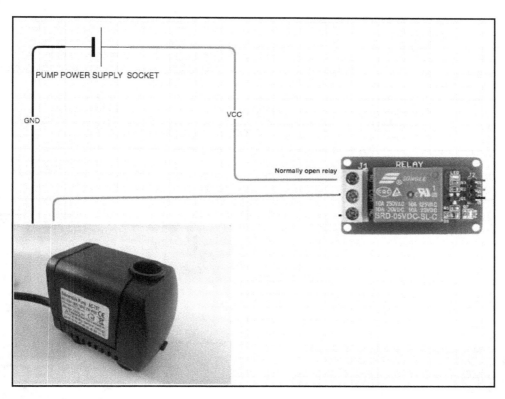

Connecting the relay to the control circuit (Arduino Pro Mini)

Now that the submersible water pump on and off status is controlled by our normally open relay, we will connect our Arduino to the relay to control it and turn the pump on as per our soil moisture readings when they indicate dry soil.

Connecting Arduino to the relay

First we will connect the Arduino, the relay, and 5V power source. If we use a USB power bank as the 5V power source, we will connect a USB cable to it and then strip the other end of the wire. You will see a red cable and black cable among other wires once you strip the other end of the USB cable. The red cable is the V+ and black cable is the GND.

The relay input has three pins(for a single channel relay). There is a V+, a GND, and an IN pin. We will use pin 9 on the Arduino to apply a voltage to the IN pin of the relay when we want to turn on the pump:

1. Connect V+ on the 5V power source to VCC of the Arduino Pro Mini.
2. Connect GND on the 5V power source to GND on the Arduino Pro Mini.
3. Connect V+ on the 5V power source to VCC of the relay.
4. Connect GND on the 5V power source to GND on the relay.
5. Connect IN on the relay to pin 9 on the Arduino Pro Mini:

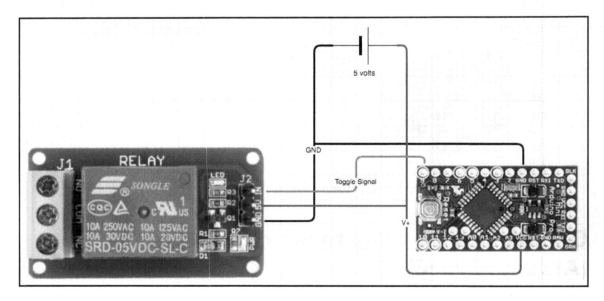

Uploading the relay test sketch to the Arduino

The following sketch will turn the relay on and off every 5 seconds:

```
/* Gardening system
 * Lets water our plants on time
```

```
 * Relay test
 */

#define RELAY_OUTPUT 9

int i = 0;

void setup() {
  // put your setup code here, to run once:
  pinMode(RELAY_OUTPUT, OUTPUT); // Set the pin to digital output mode
}

void loop() {
  // put your main code here, to run repeatedly:
  if(i == 0) {
    i = 1;
    digitalWrite(RELAY_OUTPUT,HIGH);
  } else {
    i = 0;
    digitalWrite(RELAY_OUTPUT,LOW);
  }

  delay(5000);
}
```

Upload this sketch to the Arduino Pro Mini using the FTDI cable with the Arduino IDE and once the sketch is uploaded, use a multimeter to check that the relay connects the circuit every 5 seconds.

 Ensure that the pump power is switched off before uploading this sketch or you may risk damaging your pump without water when the relay switches the pump on every 5 seconds.

Triggering the pump with the hygrometer

Now we need to update our sketch so that we trigger the relay for 5 minutes when the soil humidity sensor reads dry soil. Here is the updated sketch:

```
/* Gardening system
 * Lets water our plants on time
 */

// humidity sensor is at analog port 0
const int SENSOR_PORT = A0;
```

```
#define RELAY_OUTPUT 9

int sensorValue = 0;

void setup() {
  // We will initialize the serial port so we can print humidity data to
the serial port.
  Serial.begin(9600);
  pinMode(RELAY_OUTPUT, OUTPUT); // Set the pin to digital output mode
  digitalWrite(RELAY_OUTPUT,LOW);
}

// the loop routine runs over and over again forever:
void loop() {
  // Read the input on analog pin 0:
  sensorValue = analogRead(SENSOR_PORT);
  // A well watered soil will give a reading of atleast 380 to 400, a fully
moist soil will have reading of 1023.
  sensorValue = constrain(sensorValue, 400, 1023);
  // We will reverse the range so that we get a rough humidity percentage
reading i.e.
  // fully moist soil will give a reading of 100% for 100% humidity and
  // dry soil will read 0% for 0% humidity
  sensorValue = map(sensorValue,400,1023,100,0);

  // Print humidity value
  Serial.println(sensorValue);
  if (sensorValue <= 20) {
    Serial.println("soil is dry. switching on the PUMP");
    digitalWrite(RELAY_OUTPUT,HIGH);
    delay(300000);
  } else
  {
    Serial.println("soil is wet");
    digitalWrite(RELAY_OUTPUT,LOW);
  }

  delay(1000);
}
```

Upload this code to the sketch, dip the pump in the water reservoir, power everything up, and test this setup by dipping the hygrometer electrodes in water and taking them out to check that the pump triggers when the electrode is in the air (low humidity) and turns off when the electrodes are in water (high humidity).

Adding another sensor (photoresistor) to optimize your gardening

We are almost set. However, it is generally considered a bad idea to water the plants when there is no sunlight, that is, at night. This is because photosynthesis occurs during the day and that's when plants need water the most. So we will add a photoresistor to detect the ambient light levels to optimize our plant-watering intervals.

Connecting the photoresistor

Place the photoresistor in an area where ambient sunlight would fall on it at all times of the day. Connect the photoresistor as follows:

1. Connect one pin of the photoresistor to VCC of the Arduino Pro Mini.
2. Connect the other pin of the photoresistor to pin A1 of the Arduino Pro Mini.

Updating the sketch to add photoresistor readings to the decision-making

Now we will update our sketch to ensure that we only water the plants when there is enough sunlight:

```
/* Gardening system
 * Lets water our plants on time
 */

// humidity sensor is at analog port 0
const int SENSOR_PORT = A0;
const int PHOTO_RESISTOR = A1;
#define RELAY_OUTPUT 9

int sensorValue = 0;

void setup() {
  // We will initialize the serial port so we can print humidity data to
the serial port.
  Serial.begin(9600);
  pinMode(RELAY_OUTPUT, OUTPUT); // Set the pin to digital output mode
  digitalWrite(RELAY_OUTPUT,LOW);
}

// the loop routine runs over and over again forever:
```

```
void loop() {
  // Read the input on analog pin 0:
  sensorValue = analogRead(SENSOR_PORT);
  // A well watered soil will give a reading of atleast 380 to 400, a fully
moist soil will have reading of 1023.
  sensorValue = constrain(sensorValue, 400, 1023);
  // We will reverse the range so that we get a rough humidity percentage
reading i.e.
  // fully moist soil will give a reading of 100% for 100% humidity and
  // dry soil will read 0% for 0% humidity
  sensorValue = map(sensorValue,400,1023,100,0);

  // Print humidity value
  Serial.println(sensorValue);

  // Ensure that there is some ambient light before turning the pump on
  if (sensorValue <= 20 && analogRead(PHOTO_RESISTOR) > 25) {
    Serial.println("soil is dry. switching on the PUMP");
    digitalWrite(RELAY_OUTPUT,HIGH);
    delay(300000);
  } else
  {
    Serial.println("soil is wet");
    digitalWrite(RELAY_OUTPUT,LOW);
  }

  delay(1000);
}
```

This sketch reads the photoresistor and uses that as an additional condition to ensure that your plants are only watered during the day. We have set the light threshold as 25, but you may try to experiment with different light values to see what works best for you.

Connecting it all together

Now all that is left is for you is to put your connected electronics (5V power supply, the Arduino Pro mini, and the relay) in the junction box and stick them using a hot-melt glue gun.

You may also use the hot-melt glue to waterproof any other parts that may need waterproofing:

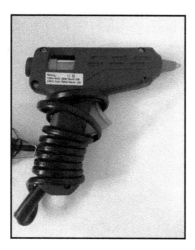

After placing the Arduino, relay, and the 5V power supply safely in the junction box, connect your drip irrigation system to the submersible pump and submerge the connected pump in your water reservoir. Place your drip irrigation system to directly water the roots of your plants and finally switch on the pump power supply.

Summary

This is it! You now have a fully automated gardening system. You can experiment with a few more things; for example, you may try connecting potentiometer knobs to seasonally adjust the desired soil moisture settings for your plants as needed!

For a more advanced gardening system, you may also try to hook up a soil acidity level sensor and have an additional setup to balance the soil pH levels as per your plant's requirements. Don't let your lack of a green thumb hold you back from getting a garden that your neighbor would be envious of!

In the next chapter, we will look at how to create a snazzy IoT blinky lights festival display using an open source project that lets the world collaborate to color your lights!

3
Integrating CheerLights into a Holiday Display

CheerLights is a project by Hans Charler that allows your lights to be controlled by Twitter tweets. Anytime somebody tweets to twitter to `@CheerLights` with a color, all the CheerLights around the world will change to that color. This section will help you brush up on your soldering skills, navigate portions of the ESP8266 breakout board, and encourage creative building of LEDs into a variety of items.

In this chapter, we will create a festive holiday display that is triggered by festive cheers sent to you by your friends.

We will discuss the following topics:

- Items required for this project
- Getting the CheerLights code set up
- Connecting it all together
- Programming the ESP8266 Huzzah for CheerLights

Items required for this project

You will need the following materials for this project:

- Adafruit HUZZAH ESP8266 breakout
- One USB FTDI connector
- Two WS2811 NeoPixels LED strips

- Two 5V and 3.3V (optional) power supplies
- Two 1000μF capacitors
- Two logic level shifting ICs
- Three 400-600Ω resistor 3 JST SM or JST PH connectors (optional)

Adafruit HUZZAH ESP8266 breakout

Adafruit Huzzah ESP8266 is a Wi-Fi microcontroller based on the Arduino IDE and it's one of the many ways in which you can create a CheerLights display. Note that you may choose to create a CheerLights display in scores of different controller types; however, for the purpose of this chapter, we will be using the Adafruit Huzzah ESP8266 for the convenience of included Wi-Fi. It also comes preprogrammed with the `NodeMCU` library, but we will be flashing it using the Arduino IDE with our own stuff. If you need to get the Lua interpreter back afterward, use the flasher to reinstall it.

USB FTDI connector

You will need a USB FTDI connector to connect the ESP8266 to your computer. This is required for you to program your ESP8266. One option to purchase this cable is at Adafruit. You can find it at `https://www.adafruit.com/product/70`.

WS2811 NeoPixels LED strip

WS2811 LEDs are individually addressable LEDs which we will use for lighting up our display.

5V and 3.3V (optional) power supply

The ESP8266 works on a 3.3V power supply while the LEDs use 5V. You can choose to power them with separate power supplies, but if you prefer to use a single power supply, you must get a 5V power supply and DC/DC converter or voltage regulator (with a minimum of 250mA output) to step down from 5V to 3.3V for the ESP8266. The 3.3V 250mA linear voltage regulator—L4931-3.3 TO-92 (`https://www.adafruit.com/product/2166`) would do nicely for this task. In our example we will use the L4931-3.3 TO-92 voltage regulator to get the 3.3V output.

Then, make sure that your power supply provides enough amperage depending on the number and rating of the LEDs you intend to use. If you use a separate power supply for the ESP8266 and NeoPixels, make sure that you connect the ground on both the power supplies together.

The ESP8266 consumes less than 250mA of power and the power supply will generally supply a few amps of power, so your primary power consumer will be the LEDs that you use. Calculate the power adapter you need accordingly. A good choice is the 5V, 10 A adapter from Adafruit (`https://www.adafruit.com/product/658`), which will be enough to power a 5-meter long strip of Adafruit NeoPixels LEDs with 32 LEDs per meter. Each NeoPixel draws 60mA at full brightness (`https://www.adafruit.com/product/1138`).

1000µF capacitor

It is recommended you get a 1000µF or higher capacitor to put in parallel with the LED strip in order to avoid damaging the LEDs from the sudden onset of current when the power source is switched on. The capacitor adds a buffer for sudden draws of power from the LEDs.

Logic level shifting IC

The ESP8266 works on a 3.3 V voltage level while the LEDs have a voltage rating of 5 volts, so you will need a level-shifting IC to switch the voltage from 3.3 volts to 5 volts for the connection between the LEDs and the ESP8266. The 74AHCT125—a quad level shifter from Adafruit (`https://www.adafruit.com/product/1787`)—is one viable choice for our purpose, and this is what we will use in our examples.

400-600Ω resistor

It is recommended you slip in a 400-600Ω resistor in between the DIN and the ESP8266. This prevents sudden spikes on the data line from damaging your pixel.

JST SM or JST PH connectors (optional)

It's generally a good idea to have connectors between your LED strip and everything else for convenient and safe storage of your lights when you are not using them. If your LEDs come with a connector, you may use that, or else you may solder on a JST SM connector to the LED strip.

Getting the CheerLights code set up

In this section, we will set up our CheerLights code and deploy it to the ESP8266 Huzzah.

Set up the Arduino IDE

The Arduino IDE is a free IDE provided by `https://www.arduino.cc/` that makes it easy to write and deploy code on Arduino and Arduino-based boards like the ESP8266. It's written in Java and it supports Windows, GNU/LINUX, and macOS. Before we start, you must install the Arduino IDE from the following link. You might have to select the correct version for your OS.

`https://www.arduino.cc/en/main/software`

If you already have the Arduino IDE, make sure its version is higher than 1.6.4.

By default, the Arduino IDE doesn't support the ESP8266. The Arduino IDE needs to be set up to work with the ESP8266; to do that, follow these steps:

1. Open up your Arduino IDE.
2. Under the menu, look for and open **Preferences**.
3. In the input box for **Additional Boards Manager URLs**, enter this URL: `http://arduino.esp8266.com/stable/package_esp8266com_index.json`.
4. Click on **OK**. Open **Boards Manager** at **Tools** | **Board:*** | **Boards Manager**.
5. Under **Type**, select **Contributed** and type `esp8266` in the search bar.
6. Click on **Install** to install support for the ESP8266.

These instructions are for Arduino IDE 1.8.1 and may differ depending on your version of IDE.

Once you have installed the ESP8266, you may have to restart the IDE. After restarting the Adafruit Huzzah, the ESP8266 will be visible under **Tools | Board:***. When you select your board, follow these steps:

1. Select **80MHz** in **Tools | CPU Frequency**.
2. Select **115200 baud** in **Tools | Upload Speed**.

Once this setup is done, connect the ESP8266 to the USB port on your computer using a USB FTDI cable.

Installing the NeoPixel library

The Adafruit NeoPixel library is the Arduino library for controlling the NeoPixel's LED strip.

In the Arduino IDE menu, go to **Sketch | Include Library | Manage Libraries...** and search for the `Adafruit NeoPixel` library by Adafruit. Install the library and restart your IDE.

Testing the ESP8266

We can test the ESP8266 by running the standard Arduino blink test on it. In the Arduino IDE, click on **New** and paste the following default Arduino blink code in it:

```
// The setup function runs once when you press reset or power the board
void setup() {
    // initialize digital pin LED_BUILTIN as an output.
    pinMode(LED_BUILTIN, OUTPUT);
}

// The loop function runs over and over again forever
void loop() {
    digitalWrite(LED_BUILTIN, HIGH);    // turn the LED on
    delay(1000);                        // wait for a second
    digitalWrite(LED_BUILTIN, LOW);     // turn the LED off
    delay(1000);                        // wait for a second
}
```

Connect the ESP8266 Huzzah to the USB FTDI connector (FT232RL) and your computer

Connect the ESP8266 to the USB FTDI connector (FT232RL):

1. Connect your **GND** on the ESP8266 Huzzah to the **GND** on your USB FTDI connector.
2. Connect the **V+** on the ESP8266 Huzzah to **V+** (or VCC) on your USB FTDI connector.
3. Connect **Tx** on the ESP8266 Huzzah to **Rx** on your USB FTDI connector.
4. Connect **Rx** on the ESP8266 Huzzah to **Tx** on your USB FTDI connector.
5. Now connect the USB FTDI connector (FT232RL) to a USB port on your computer:

Switching the ESP8266 Huzzah into deployment mode

To bring the ESP8266 Huzzah into deployment mode, you must go through the following steps:

1. Press and hold down the **GPIO0** button on your ESP8266 Huzzah. The red LED on your board will light up.
2. While holding down the **GPIO0** button, press the **RESET** button on your ESP8266 Huzzah.
3. After a second, release the **RESET** button. The red LED on the board will go dimmer.

The board is now in deployment mode and ready for you to push code. There is no need to rush at this moment as there is no time limit for you to upload your code.

 The ESP8266 Huzzah must be switched into deployment mode using these steps every time you need to upload an Arduino sketch to it. Make sure you follow these steps in the next section before you upload your CheerLights code.

Deploying your code

Deploy this sketch on your ESP8266 using the deploy button, which is usually an arrow on the new versions of the Arduino IDE.

If the LED light on the ESP8266 blinks with a 1-second period between blinks, that means your board is working just fine.

Connecting it all together

Adafruit the NeoPixels LED strip has three inputs: GND, V+ (or **VIN**) and DIN (or **Digital Input**). The ordering of the inputs may differ depending on when and who you order the strip from. Follow these steps:

1. First connect the capacitor and power supply in parallel.
2. Connect the **1000µF** capacitor in parallel across the **5V** power supply. If your multimeter supports capacitance measurement and **1000µF** is within the measuring range, you may use it to verify the capacitance of your capacitor.

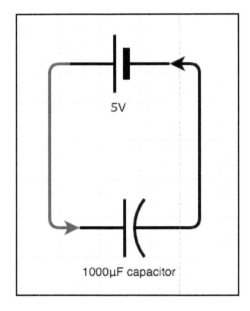

3. Connect up the grounds next. Connect the power supply ground to the LED strip ground, to the 74AHCT125 ground and the **10E** pins, and to ground on the ESP8266. If you have a **3.3V** supply, connect the grounds of the two power supplies as well.

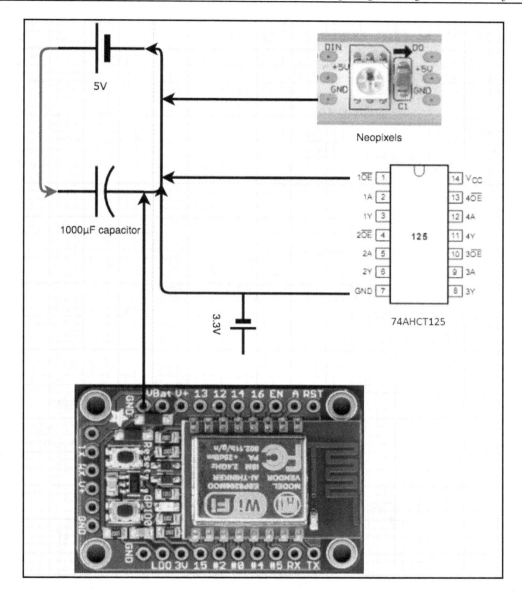

Connecting the data pins

Now we must go ahead and connect the data pins. These pins are used for communication between the ESP8266 Huzzah and the NeoPixel strip. Since NeoPixels communicate on digital 5V signal and the ESP8266 communicates on a 3.3V digital signal, we add a 74AHCT125 IC as an intermediary to shift signal voltage levels:

1. Connect pin **13** from the ESP8266 to **1A** on the **74AHCT125**.
2. Finally connect the **DIN** of the LED to **1Y** on the **74AHCT125**.

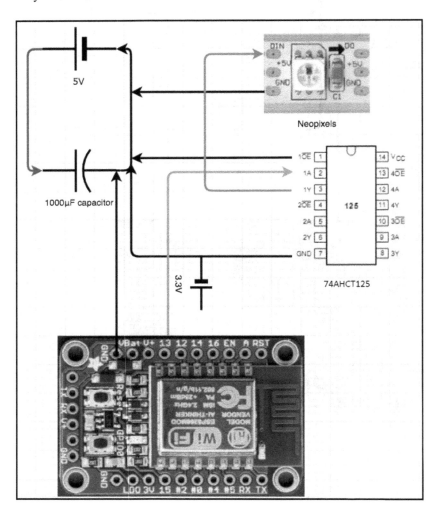

Power the NeoPixels

Connect the power supply's **5V** terminal to the **VCC** (or V+) on the NeoPixels strip and to **VCC** (or V+) on the **74AHCT125**:

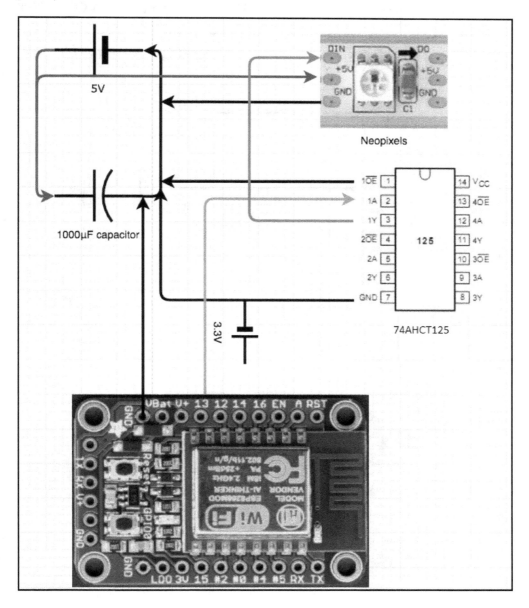

Power up the ESP8266

If you have a DC/DC converter or voltage regulator to get the **3.3V** for the ESP8266, then connect the power supply's **5V** terminal to the ESP8266 bridging through the **5V** to **3.3V** DC/DC converter or connect the **3.3V** from the **3.3V** power supply to the **VCC** of ESP8266:

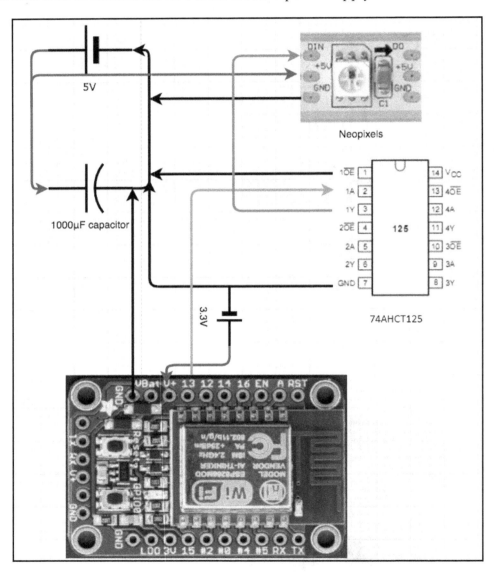

If you are using two power supplies (5V and 3.3V) make sure you connect their grounds together.

 Be careful to connect the **-ve** on the capacitor to the ground and **+ve** to the **+ve** on the power supply. If you connect it in reverse, the capacitor will blow up. Most capacitors have a dash symbol to mark the negative wire. The positive leg is often longer than the negative leg to mark the difference.

Programming the ESP8266 Huzzah for CheerLights

In this section, we will learn how to program the ESP8266 Huzzah to link up with the CheerLights API and talk to NeoPixels.

Let's try a simple single-color display

We are almost done with our funky networked blinking lights now; all we need to do is program the ESP8266 with the CheerLights code. For this, connect the ESP8266 with your computer again and switch back to the Arduino IDE:

1. Include the libraries: The first thing we need to do in the code is include the libraries we will need. We need the Adafruit NeoPixel library and `ESP8266WiFi` library. Start with including these libraries:

    ```
    /* CheerLights
     * Let the world set your LED strip's color using twitter.
     */

    #include <ESP8266WiFi.h>
    #include <Adafruit_NeoPixel.h>
    ```

2. Preprocessor directives: Since we have connected the NeoPixels (via level shifter) to pin **13** on the ESP8266 Huzzah, we must define that as well. So add the following to your code:

    ```
    #define NUMBER_OF_LEDS 50  // Number of LEDs on your NeoPixels
    strip

    #define HUZZAH_DATA_PIN 13 // The ESP8266 Pin on which your LEDs
    ```

```
                                           // are connected
```

3. Wi-Fi configuration: Set the Wi-Fi SSID (access point name) and password:

```
const char* ssid = "MyWifi";
const char* password = "WifiPassword";
```

4. CheerLights API endpoint: Specify the API endpoint:

```
// This is the thingspeak CheerLights api endpoint details.
const char* apiHost = "api.thingspeak.com";
const char* apiUrl = "/channels/1417/field/2/last.txt";
long apiLastPolledTime = 0;
```

5. Color values: These variables will store the current color retrieved from the CheerLights API:

```
uint8_t cheerLightsRed;
uint8_t cheerLightsGreen;
uint8_t cheerLightsBlue;
```

6. NeoPixels `strip` object: Define your NeoPixels `strip` object. Exact values might differ based on your NeoPixels strip. Refer to NeoPixel documentation if necessary:

```
Adafruit_NeoPixel strip =
Adafruit_NeoPixel(NUMBER_OF_LEDS,HUZZAH_DATA_PIN, NEO_RGB + NEO_KHZ400);
```

7. The `setup()` method: Define `setup()` for pre-initialization:

```
void setup() {
    // Connecting to the Wifi for the first time
    Serial.begin(115200);
    delay(100);

    setupWifi();
    strip.begin();
    strip.show();
}
```

8. The `loop()` method: Now the fun begins; let's define the main loop that will do the meat of the work for us:

```
void loop() {
    // Update color every three seconds
    if(millis() - apiLastPolledTime > 3000) {
        updateCheerLightsColor();
```

```
                    Serial.print("Received color: " + (String)cheerLightsRed +
    ", " + (String)cheerLightsGreen + ", " + (String)cheerLightsBlue);
        }

            // Set NeoPixels color
            setNeoPixelsColor();
            delay(1000);
        }
```

9. The `setupWifi()` function: Define the `setupWifi` function for connecting to the Wi-Fi:

```
    void setupWifi() {
        Serial.printf("Attempting to connect to Wifi: %s\n", ssid);
        WiFi.begin(ssid, password);
        while (WiFi.status() != WL_CONNECTED) {
            Serial.print(". ");
            delay(1000);
        }
        Serial.println("");
        Serial.println("WiFi connected");
        Serial.println("IP address: ");
        Serial.println(WiFi.localIP());
        Serial.print("Netmask: ");
        Serial.println(WiFi.subnetMask());
        Serial.print("Gateway: ");
        Serial.println(WiFi.gatewayIP());
    }
```

10. The `updateCheerLightsColor()` method: Define `updateCheerLightsColor()`. This function gets the current CheerLights color from the twitter API:

```
    void updateCheerLightsColor() {
        // Get the TCP client object
        WiFiClient client;
        const int httpPort = 80;
        if (!client.connect(apiHost, httpPort)) {
            Serial.println("Unable to connect to the api endpoint");
            return;
        }

        // Send the HTTP request with a GET Header
        client.print(String("GET ") + apiUrl + " HTTP/1.1\r\n" +
                "Host: " + apiHost + "\r\n" +
                "Connection: close\r\n\r\n");
```

```
                delay(1000);

                // Listen to the HTTP Response and parse it line by line
                // till your reach the response body(color) line(The one
                // that starts with a #.
                String line = "";
                while(client.available()){
                    line = client.readStringUntil('\r');
                    // Thinkspeak api returns a HTML hex color when it works.
                    // We check if there is a hex color by looking for a
hash(#)
                    // in a line in the response body.
                    int hashindex = line.indexOf('#');
                    if (line.indexOf('#') > 0) {
                        // Get the position of # in the line and get the color
                        // values next to it.
                        String colorhex = line.substring(hashindex,
hashindex+7);
                        long color = (int) strtol( &colorhex[1], NULL, 16); //
to RGB

                        // Extract single byte value for each color and put it
in global
                        // variables
                        cheerLightsRed = color >> 16;
                        cheerLightsGreen = color >> 8 & 0xFF;
                        cheerLightsBlue = color & 0xFF;
                    }
                }
                apiLastPolledTime = millis();
            }
```

11. The `setNeoPixelsColor()` function: We will start off with a simple
    `setNeoPixelsColor()` that loops through all NeoPixels and updates them to
    the latest color received from the last CheerLights API call available in the
    variables `cheerLightsRed`, `cheerLightsGreen`, and `cheerLightsBlue`:

```
        void setNeoPixelsColor() {
            for (uint16_t i = 0; i < strip.numPixels(); i++) {
                strip.setPixelColor(i, cheerLightsRed, cheerLightsGreen,
cheerLightsBlue);
                strip.show();
            }
        }
```

Deploy this sketch on your ESP8266 Huzzah using the deploy button, which is usually an
arrow on the new versions of the Arduino IDE: .

Make sure you switch your Huzzah to deployment mode using the steps outlined earlier before you click on deploy.

If you are reading the serial port on the Huzzah, you will start seeing the messages as the board first connects to Wi-Fi and starts reading the CheerLights API endpoint.

It's time to start tweeting now. You may tweet something like:

```
@CheerLights Paint the town red
```

And the LED will change to red within 3 seconds (provided no one else tweets some other color before your CheerLights update).

Let's try some interesting modifications

In the previous section, we just created a display that instantly swaps all pixel colors on any CheerLights update.

NeoPixels are individually addressable LEDs, which means you can create a lot more interesting display with your CheerLights. Let's try to create a displays that slowly updates all your pixels gradually as the color changes in a fluid motion. To do this, update the code in your `setNeoPixelsColor()` function to the following:

```
void setNeoPixelsColor() {
    for (uint16_t i = 0; i < strip.numPixels(); i++) {
        delay(100);
        strip.setPixelColor(i, cheerLightsRed, cheerLightsGreen,
cheerLightsBlue);
        strip.show();
    }
}
```

This will create a flowing motion update of color every time there is a color change.

You may try other interesting combinations of displays. As an exercise, try to make a shimmering effect if the color is not updated in a while.

Summary

CheerLights is a fun and simple collaborative project to start with in your home automation journey. In this chapter, we also learned how to control NeoPixels, which are individually addressable LEDs. What you can do with the NeoPixels is only limited by your imagination. The NeoPixels are ridiculously easy to control and they can be used for some interesting automation projects. Some more practical applications can be to indicate system status of server racks on your workstation desk or to create floating ticker boards. Let your imagination run wild and share what you create!

In the next chapter, we are going to learn how to create a parking space monitor to automatically alert you when your designated parking space is occupied using a machine learning system to ease your vehicle parking headaches. Excited? Read on to learn more.

Cheers!

4

Erase Parking Headaches with OpenCV and Raspberry Pi

Searching for a parking space is a routine (and often frustrating) activity for many people in cities around the world. In big cities after arriving in rush hour at a mall or somewhere, parking takes a long time. It's inefficient, frustrating, and time-consuming when we need to park our car. It's a big problem to find a parking space, and in this chapter we will see how to build a smart parking lot that sends notifications and informs someone can occupy that vacant place with the internet of the cities, using an automated vision system with a Raspberry Pi 3 board.

In this chapter we will learn the following topics:

- Introduction to smart parking systems
- Sensor devices for smart parking
- Vision machine systems
- Logging spaces in a database
- Detecting parking space
- Amazon Web Services IoT
- Integrating the system

Let's get started!

Introduction to smart parking systems

It's a problem in big cities to get parking: searching for a vacant parking space takes a long time. The problem is growing since the number of cars is enormous. In a general parking architecture, there are some areas that we can consider will be present:

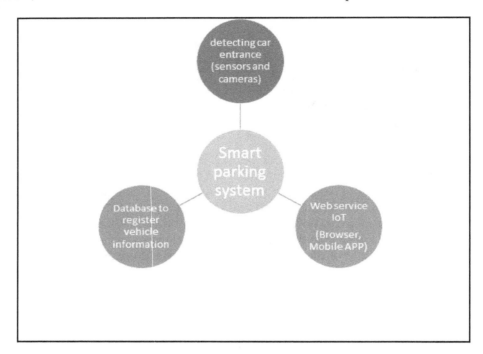

- **Detecting car entrance**: This is the hardware to detect when the car enters the parking lot
- **Database to register vehicle information**: Registers the information for the cars; this will help get complete information about the parking lot
- **Web services IoT**: Devices connected to the internet get information from the main system, and display the data on a dashboard in real time

Sensor devices for smart parking

To build a smart parking system, it's necessary to have some sensors that can be applied to such a system. Let's see them in detail.

Presence sensor

To detect when a car is in a parking lot, there are some sensors available; in this case, we will use a presence sensor like this:

In some parking lots, there is a gate that goes up when the driver arrives in front of it and they press the button and receive their ticket; in others, they have an identity card. Here is a common parking lot:

This is a prototype of the entrance to the parking lot:

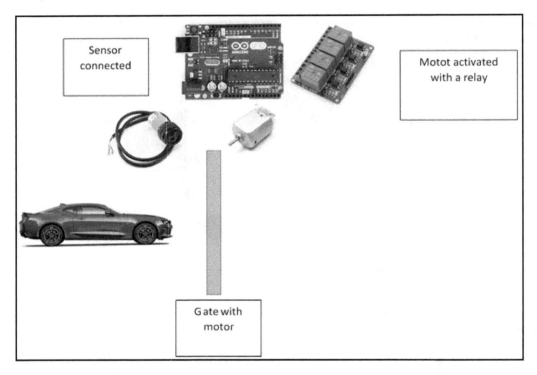

Now we will make a sketch that we can use with Arduino; it detects when the car gets close to the sensor. We use the digitalread() instruction to detect a digital signal; this sensor activates with a signal in low, as shown in the following code:

```
#include <Wire.h>
#include <LiquidCrystal_I2C.h>

LiquidCrystal_I2C lcd(0x3F,16,2); // set the LCD address to 0x20 for a
16 chars and 2 line display

//variables

int cont = 0;
int reading;

void setup()
{
  Serial.begin(9600);
  lcd.init();
```

```
    pinMode(13,OUTPUT);
    pinMode(2,INPUT_PULLUP);
    // Print a message to the LCD.
    lcd.backlight();
    lcd.print("Counting Cars");
    lcd.setCursor(0,1);
    lcd.print("Counting = 0 ");
  }

void loop()
{
  lectura = digitalRead(2);
  if (reading == LOW){
    digitalWrite(13,HIGH);
    delay(1000);
    digitalWrite(13,LOW);
    cont = cont + 1;
    lcd.setCursor(0,1);
    lcd.print("Counting = ");
    lcd.print(cont);
    Serial.print("Counting = ");
    Serial.println(cont);

    // this condition counts until 21 cars after that the count returns
to zero
    if (cont == 21){
      cont = 0;
      lcd.setCursor(0,1);
      lcd.print("  ");
      lcd.setCursor(0,1);
      lcd.print("Counting = ");
      lcd.print(cont);
      Serial.print("Counting = ");
      Serial.println(cont);
    }
  }
}
```

The final connections look like this:

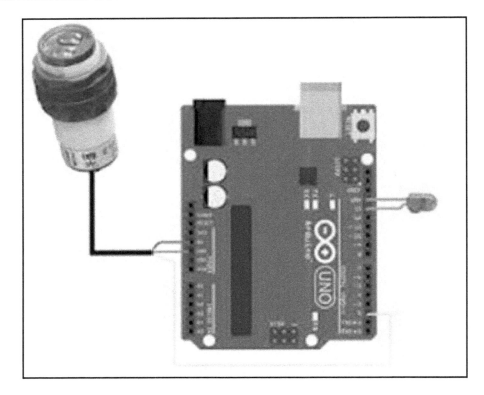

This information of the total number can be displayed on a big screen or we implement storing the number of cars or spaces in a database; we will discuss this in the following sections.

Ultrasonic sensor

The HC-SR04 is a cheap ultrasonic sensor. It is used to measure the range between itself and an object. Each HC-SR04 module includes an ultrasonic transmitter, a receiver, and a control circuit. You can see that the HC-SR04 has four pins: GND, VCC, Trigger, and Echo. You can buy HC-SR04 from SparkFun, at `https://www.sparkfun.com/products/13959`, as shown in the next image. You can also find this sensor on Seeed Studio at `https://www.seeedstudio.com/Ultra-Sonic-range- measurement-module-p-626.html`:

To work with the HC-SR04 module, we can use the NewPing library on our
Sketch program. You can download it from `http://playground.arduino.cc/Code/NewPing`
and then deploy it into the Arduino `libraries` folder. After it is deployed, you can start
writing your Sketch program.

Now open a Sketch and write the following code:

```
#include <NewPing.h>

#define TRIGGER_PIN 2
#define ECHO_PIN 4
#define MAX_DISTANCE 600

NewPing sonar(TRIGGER_PIN, ECHO_PIN, MAX_DISTANCE);

long duration, distance;

void setup() {
  pinMode(13, OUTPUT);
  pinMode(TRIGGER_PIN, OUTPUT);
  pinMode(ECHO_PIN, INPUT);
  Serial.begin(9600);
}

void loop() {
  digitalWrite(TRIGGER_PIN, LOW);
  delayMicroseconds(2);

  digitalWrite(TRIGGER_PIN, HIGH);
```

```
      delayMicroseconds(10);

      digitalWrite(TRIGGER_PIN, LOW);
      duration = pulseIn(ECHO_PIN, HIGH);

      //Calculate the distance (in cm) based on the speed of sound.
      distance = duration/58.2;

      Serial.print("Distance=");
      Serial.println(distance);
      delay(200);
}
```

Camera

A camera will be used in this project; this is a sensor for image processing and machine learning systems. We will use in this case a camera for the Raspberry Pi. We recommend the camera at `https://www.raspberrypi.org/products/camera-module-v2/`:

Configure the camera

This camera module is connected to the Raspberry Pi through the CSI interface. To use this camera, you should activate it in your OS. For instance, if we use the Raspbian OS, we can activate the camera from the command line. You can type this command:

```
sudo apt-get update
sudo raspi-config
```

We should see the following menu to enable the camera on the Raspberry Pi:

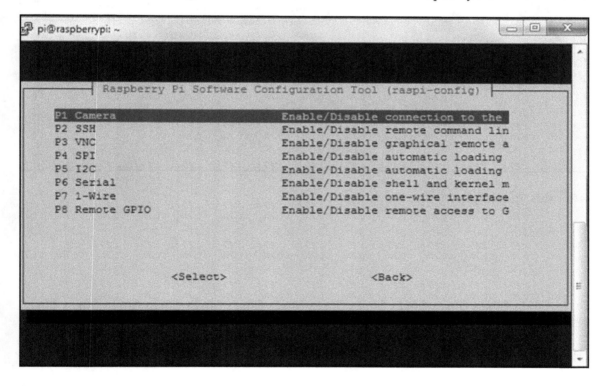

Select enable camera from the menu:

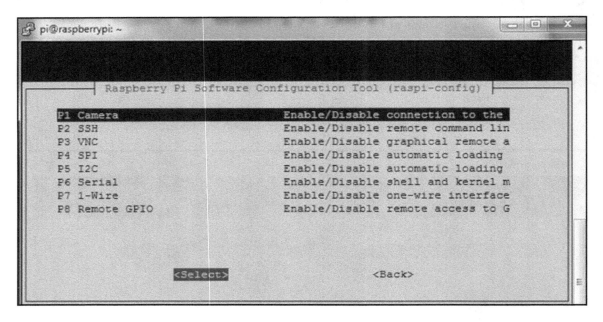

After this, the Raspberry Pi will reboot.

Now that everything is connected, we can see camera connected to the Raspberry Pi in the following image:

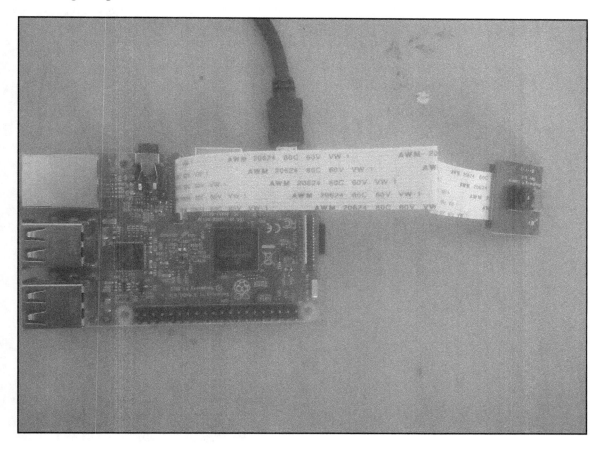

Accessing our Raspberry Pi via SSH

When we want to access our Raspberry Pi, we need to get its IP address. To get it, it's necessary to connect it to a router via an Ethernet cable, configure the router to assign DHCP, and write the IP address that we will use, as shown in the following screenshot:

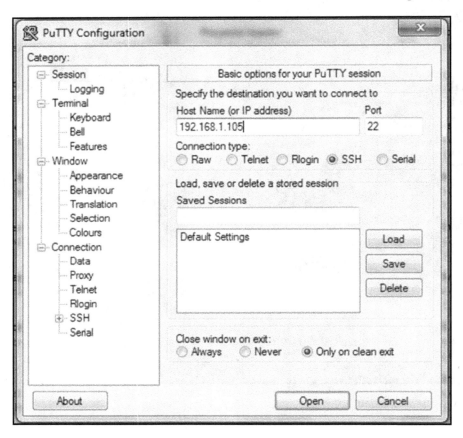

 It's recommended you create a file with the name SSH in the root directory of the SD card; this will help you connect with PuTTY and create remote access.

After this, the system will require the username and password; in this case, they are `pi` and `raspberry`, respectively:

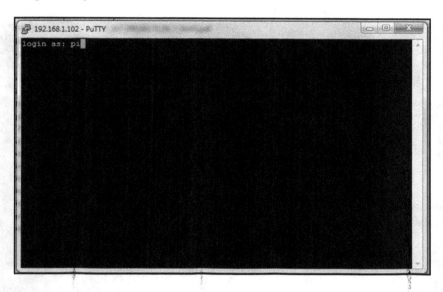

Here, enter the password:

The last screen shows the Pi is connected and ready to receive commands:

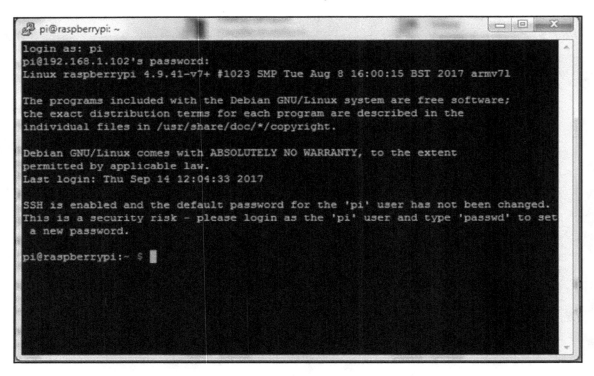

Machine vision systems

In this section, we will see some concepts about machine vision, the requirements to understand detection, and processing the image to be applied to the machine vision system.

Introduction to machine vision

Machine vision is a machine with camera capabilities and an understanding of what objects are. The machine uses its camera to sense physical objects around its environment. Machine vision or computer vision is a field where a machine acquires, analyzes, and understands a still image or video. This field involves techniques such as image processing, pattern recognition, and machine learning.

The pattern recognition and machine learning fields help us teach our machine to understand images. For instance, when we show a still image with people inside a car to the machine, then the machine should identify the people. Furthermore, in some cases the machine also should guess the person in an image. From a pattern recognition and machine learning viewpoint, we should register the person so the machine can know the person in the image after identifying the person in an existing image. To build a machine vision system, we use the general design that is shown in the following figure:

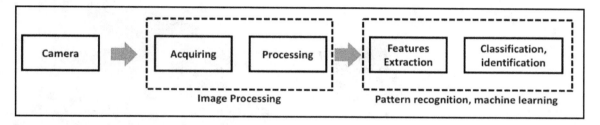

Firstly, we acquire images from a **Camera**. Each image will be processed for **Image Processing** tasks such as removing noise, filtering, or transforming. Then, we perform **Feature Extraction** for each image. There are various feature extraction techniques depending on your purposes. After obtaining the features of images, we identity and recognize objects in an image. **Pattern recognition** and **machine learning** are involved in this process.

Testing the camera

We verify the camera has been detected on the **Camera Serial Interface** (**CSI**) interface, using the following command:

```
vcgencmd get_camera
```

Now we see the camera is responding in the next image:

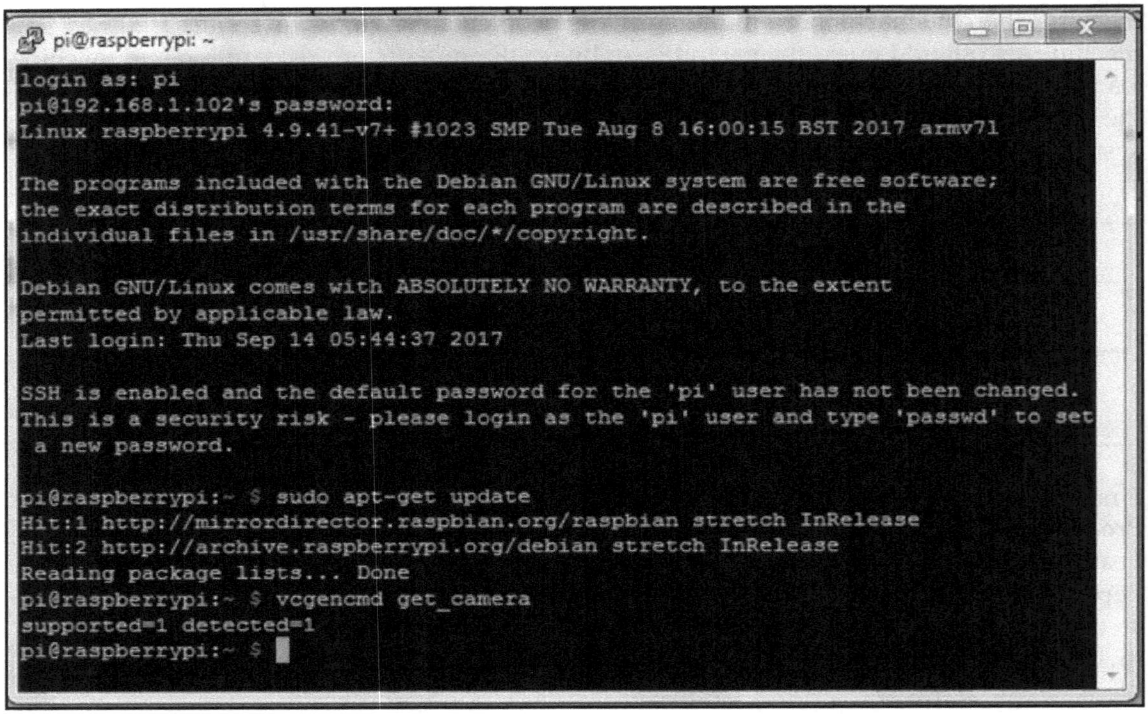

For testing, we'll develop a Python application to take a photo and save it to local storage. We'll use the `PiCamera` object to access the camera.

We use the following script:

```
from picamera import PiCamera
from time import sleep
camera = PiCamera()
camera.start_preview()
sleep(5)
camera.capture('/home/Pi/Documents/image.jpg')
camera.stop_preview()
print("capture is done")
```

If everything is working and succeeds, you should see the image taken by the camera:

Installing OpenCV on the Raspberry Pi

The **Open Computer Vision** library (known as **OpenCV**) is an open source library that is designed for computational efficiency and with a strong focus on real-time applications. This library is written in C/C++ and also provides several bindings for other programming languages. The official website for OpenCV is http://www.opencv.org.

The OpenCV library provides a complete library starting from basic computation and image processing to pattern recognition and machine learning. I've noticed several research papers use this library for simulation and experiments, so this library is a good point for starting our project in machine vision/computer vision. Currently, the OpenCV library is available for Windows, Linux, Mac, Android, and iOS. You can download this library at http://opencv.org/downloads.html. I'll show you how to deploy OpenCV on a Raspberry Pi with the Raspbian OS:

```
$ sudo apt-get update
$ sudo apt-get install build-essential git cmake pkg-config libgtk2.0-dev
$ sudo apt-get install python2.7-dev python3-dev
$ sudo apt-get install libjpeg-dev libtiff5-dev libjasper-dev libpng12-
dev
$ sudo apt-get install libavcodec-dev libavformat-dev libswscale-dev
libv4l-dev
$ sudo apt-get install libxvidcore-dev libx264-dev
$ sudo apt-get install libatlas-base-dev gfortran$ mkdir opencv
$ cd opencv
$ git clone https://github.com/Itseez/opencv.git
$ git clone https://github.com/Itseez/opencv_contrib.git
$ sudo pip install virtualenv virtualenvwrapper
$ sudo rm -rf ~/.cache/pip
```

Detecting the vehicle plate number

In a parking lot, the main point is to do the things automatically; in this case, we need to make it so that, using a camera we can detect the plate with a program configured in the Raspberry Pi.

OpenALPR

OpenALPR is an open source automatic license plate recognition library written in C++ with bindings in C#, Java, Node.js, Go, and Python. You can visit the official website at `https://github.com/openalpr/openalpr`.

OpenALPR includes a command-line utility. Simply typing `alpr [image file path]` is enough to get started with recognizing license plate images:

Plate	Confidence (%)	Processing Time (ms)
PE3R2X	88.94%	152.94 ms

In the following code, we analyze and test a plate number and its processed image:

```
user@linux:~/openalpr$ alpr ./samplecar.png
```

```
plate0: top 10 results -- Processing Time = 58.1879ms.
    - PE3R2X     confidence: 88.9371
    - PE32X      confidence: 78.1385
    - PE3R2      confidence: 77.5444
    - PE3R2Y     confidence: 76.1448
    - P63R2X     confidence: 72.9016
    - FE3R2X     confidence: 72.1147
    - PE32       confidence: 66.7458
    - PE32Y      confidence: 65.3462
    - P632X      confidence: 62.1031
    - P63R2      confidence: 61.5089
```

This is the command-line instruction:

```
user@linux:~/openalpr$ alpr --help
USAGE:

   alpr  [-c <country_code>] [--config <config_file>] [-n <topN>] [--seek
         <integer_ms>] [-p <pattern code>] [--clock] [-d] [-j] [--]
         [--version] [-h] <image_file_path>
Where:
   -c <country_code>,  --country <country_code>
     Country code to identify (either us for USA or eu for Europe).
     Default=us
   --config <config_file>
     Path to the openalpr.conf file
   -n <topN>,  --topn <topN>
     Max number of possible plate numbers to return.  Default=10
   --seek <integer_ms>
     Seek to the specified millisecond in a video file. Default=0
   -p <pattern code>,  --pattern <pattern code>
     Attempt to match the plate number against a plate pattern (e.g., md
     for Maryland, ca for California)
   --clock
     Measure/print the total time to process image and all plates.
     Default=off
   -d,  --detect_region
     Attempt to detect the region of the plate image.  [Experimental]
     Default=off
   -j,  --json
     Output recognition results in JSON format.  Default=off
   --,  --ignore_rest
     Ignores the rest of the labeled arguments following this flag.
   --version
```

```
   Displays version information and exits.
 -h,  --help
   Displays usage information and exits.
 <image_file_path>
   Image containing license plates
OpenAlpr Command Line Utility
```

Integrating the library

OpenALPR is written in C++ and has bindings in C#, Python, Node.js, Go, and Java. Check out this guide for examples showing how to run OpenALPR in your application: http://doc.openalpr.com/bindings.html.

Here is the API for Python:

```python
from openalpr import Alpr

alpr = Alpr("us", "/path/to/openalpr.conf", "/path/to/runtime_data")
if not alpr.is_loaded():
    print("Error loading OpenALPR")
    sys.exit(1)

alpr.set_top_n(20)
alpr.set_default_region("md")

results = alpr.recognize_file("/path/to/image.jpg")

i = 0
for plate in results['results']:
    i += 1
    print("Plate #%d" % i)
    print("   %12s %12s" % ("Plate", "Confidence"))
    for candidate in plate['candidates']:
        prefix = "-"
        if candidate['matches_template']:
            prefix = "*"

        print("  %s %12s%12f" % (prefix, candidate['plate'],
candidate['confidence']))

    # Call when completely done to release memory
    alpr.unload()
```

Programming the script for Python

We will test the Raspberry Pi with a script in Python to detect the plate and get the information using a Python Terminal. First we install the following:

- OpenCV: http://opencv.org/
- Leptonica: http://www.leptonica.org/
- Tesseract OCR: https://code.google.com/p/tesseract-ocr/
- OpenALPR: https://github.com/openalpr/openalpr

After compiling everything, we will create a simple Python script that will take a photo from a camera and process it using OpenALPR.

This is the script:

```python
import json, shlex, subprocess

class PlateReader:

    def __init__(self):
    #webcam subprocess args
    webcam_command = "fswebcam -r 640x480 -S 20 --no-banner --quiet
alpr.jpg"
    self.webcam_command_args = shlex.split(webcam_command)

    #alpr subprocess args
    alpr_command = "alpr -c eu -t hr -n 300 -j alpr.jpg"
    self.alpr_command_args = shlex.split(alpr_command)

    def webcam_subprocess(self):
    return subprocess.Popen(self.webcam_command_args,
stdout=subprocess.PIPE)

    def alpr_subprocess(self):
    return subprocess.Popen(self.alpr_command_args,
stdout=subprocess.PIPE)

    def alpr_json_results(self):
    self.webcam_subprocess().communicate()
    alpr_out, alpr_error = self.alpr_subprocess().communicate()

    if not alpr_error is None:
    return None, alpr_error
    elif "No license plates found." in alpr_out:
    return None, None
```

```
        try:
        return json.loads(alpr_out), None
        except ValueError, e:
        return None, e

        def read_plate(self):
        alpr_json, alpr_error = self.alpr_json_results()

        if not alpr_error is None:
        print alpr_error
        return

        if alpr_json is None:
        print "No results!"
        return

        results = alpr_json["results"]

        ordinal = 0
        for result in results:
        candidates = result["candidates"]

        for candidate in candidates:
        if candidate["matches_template"] == 1:
        ordinal += 1
        print "Guess {0:d}: {1:s} {2:.2f}%".format(ordinal,
candidate["plate"], candidate["confidence"])

    if __name__=="__main__":
    plate_reader = PlateReader()
    plate_reader.read_plate()
```

Cloud API

The OpenALPR Cloud API is a web-based service that analyzes images for license plates as
well as vehicle information such as make, model, and color. The Cloud API service is easy
to integrate into your application via a web-based REST service. You send image data to the
OpenALPR API, they process the data, and return JSON data describing the license plate
and vehicle. For reference, you can go to `http://www.openalpr.com/demo-image.html`.

Because the OpenALPR Cloud API is REST-based, it works with any programming
language on any operating system. You can make API calls using whatever method you
prefer.

To make integration easier, the OpenALPR Cloud API also includes permissively licensed open source client libraries in a variety of languages. The GitHub repository is available here: `https://github.com/openalpr/cloudapi`.

PHP REST API script for OpenALPR

This script lets you read the data read from the camera and receive the package of data in JSON; for more information, go to `https://github.com/stefanvangastel/openalpr-php-rest-api` for your reference.

This is the JSON with data:

```
{
    "data": {
        "version": 2,
        "data_type": "alpr_results",
        "epoch_time": -1348550864,
        "img_width": 636,
        "img_height": 358,
        "processing_time_ms": 61.695,
        "regions_of_interest": [
            {
                "x": 0,
                "y": 0,
                "width": 636,
                "height": 358
            }
        ],
        "results": [
            {
                "plate": "56ZFDL",
                "confidence": 93.35331,
                "matches_template": 0,
                "plate_index": 0,
                "region": "",
                "region_confidence": 0,
                "processing_time_ms": 15.263,
                "requested_topn": -2084378561,
                "coordinates": [
                    {
                        "x": 243,
                        "y": 225
                    },
                    {
                        "x": 396,
                        "y": 227
```

```
                    },
                    {
                        "x": 396,
                        "y": 257
                    },
                    {
                        "x": 243,
                        "y": 255
                    }
                ],
                "candidates": [
                    {
                        "plate": "56ZFDL",
                        "confidence": 93.35331,
                        "matches_template": 0
                    },
                    {
                        "plate": "S6ZFDL",
                        "confidence": 85.164574,
                        "matches_template": 0
                    },
                    {
                        "plate": "56ZF0L",
                        "confidence": 84.503777,
                        "matches_template": 0
                    },
                    {
                        "plate": "56ZFOL",
                        "confidence": 83.140244,
                        "matches_template": 0
                    },
                    {
                        "plate": "B6ZFDL",
                        "confidence": 81.62719,
                        "matches_template": 0
                    },
                    {
                        "plate": "6ZFDL",
                        "confidence": 79.280426,
                        "matches_template": 0
                    },
                    {
                        "plate": "S6ZF0L",
                        "confidence": 76.315041,
                        "matches_template": 0
                    },
                    {
                        "plate": "S6ZFOL",
```

```
                    "confidence": 74.951508,
                    "matches_template": 0
                },
                {

                    "plate": "B6ZF0L",
                    "confidence": 72.777657,
                    "matches_template": 0
                },
                {

                    "plate": "B6ZFOL",
                    "confidence": 71.414124,
                    "matches_template": 0
                }
            ]
        }
    ]
  }
}
```

Architecture of the parking system

In the following figure, we have the architecture of the parking system connected to the Raspberry Pi with Ethernet modules. The Arduino Ethernet shield is connected to the server, sending the signals from the sensor and receiving the data, saving it in the server notifying the app or a web page in a browser. In the following figure, we have the scenario where we include all the parts:

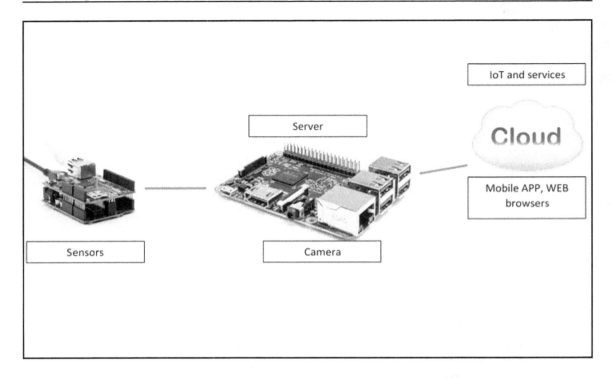

Logging spaces in a database

In this section, we will describe how to create and configure web and database servers to enable logging spaces in a parking lot using vehicle information.

Configuring a MySQL database server

In this section, you will learn how to configure a MySQL server in order to create a database and integrate everything in your dashboard in order to record data in a database.

Installing MySQL

Our Raspberry Pi 3 is being configured like a web server. In this section, we will install a MySQL database server with the following command so we can receive connections from clients, display data stored in the database, and use queries in SQL:

```
sudo apt-get install mysql-server
```

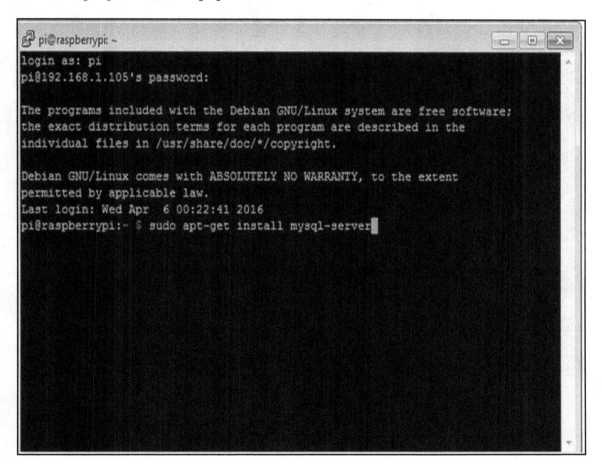

It will ask you for the password of the root user:

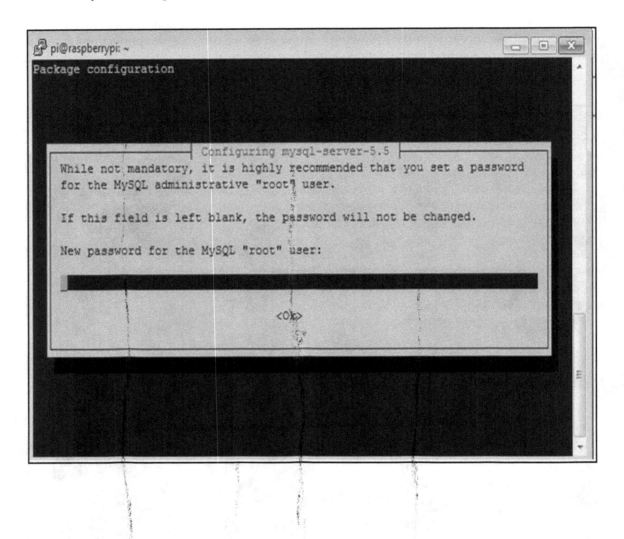

Connect to MySQL and type the following command:

```
mysql -u root -p
```

```
 pi@raspberrypi: ~
Preparing to unpack .../mysql-server-core-5.5_5.5.44-0+deb8u1_armhf.deb ...
Unpacking mysql-server-core-5.5 (5.5.44-0+deb8u1) ...
Selecting previously unselected package mysql-server-5.5.
Preparing to unpack .../mysql-server-5.5_5.5.44-0+deb8u1_armhf.deb ...
ERROR: There's not enough space in /var/lib/mysql/
dpkg: error processing archive /var/cache/apt/archives/mysql-server-5.5_5.5.44-0
+deb8u1_armhf.deb (--unpack):
 subprocess new pre-installation script returned error exit status 1
Selecting previously unselected package libhtml-template-perl.
Preparing to unpack .../libhtml-template-perl_2.95-1_all.deb ...
Unpacking libhtml-template-perl (2.95-1) ...
Selecting previously unselected package mysql-server.
Preparing to unpack .../mysql-server_5.5.44-0+deb8u1_all.deb ...
Unpacking mysql-server (5.5.44-0+deb8u1) ...
Processing triggers for man-db (2.7.0.2-5) ...
/usr/bin/mandb: can't write to /var/cache/man/1161: No space left on device
Errors were encountered while processing:
 /var/cache/apt/archives/mysql-server-5.5_5.5.44-0+deb8u1_armhf.deb
E: Sub-process /usr/bin/dpkg returned an error code (1)
pi@raspberrypi:~ $ mysql --version
mysql  Ver 14.14 Distrib 5.5.44, for debian-linux-gnu (armv7l) using readline 6.
3
pi@raspberrypi:~ $ mysql -u root -p
Enter password: 
```

Type the following command:

```
show databases;
```

Here we can see databases of the system that are now installed on the server:

```
pi@raspberrypi: ~

Aborted
pi@raspberrypi:~ $ mysql -u root -p
Enter password:
Welcome to the MySQL monitor.  Commands end with ; or \g.
Your MySQL connection id is 45
Server version: 5.5.44-0+deb8u1 (Raspbian)

Copyright (c) 2000, 2015, Oracle and/or its affiliates. All rights reserved.

Oracle is a registered trademark of Oracle Corporation and/or its
affiliates. Other names may be trademarks of their respective
owners.

Type 'help;' or '\h' for help. Type '\c' to clear the current input statement.

mysql> show databases;
+--------------------+
| Database           |
+--------------------+
| information_schema |
| mysql              |
| performance_schema |
+--------------------+
3 rows in set (0.00 sec)
```

Installing the MySQL driver for PHP

You need the MySQL driver for PHP to access the MySQL Database. Execute this command to install the PHP-MySQL driver (`mysqlnd`). It's important to have installed an Apache server on the Raspberry Pi and text the PHP for executing on the server.

It's important to install our driver to communicate from our PHP5 interface with the MySQL database server; to do that we type the following command:

```
sudo apt-get install php5 php5-mysql
```

Testing PHP and MySQL

In this section, we will make a simple page to test PHP and MySQL with the following command:

```
sudo nano /var/www/html/hellodb.php
```

```
pi@raspberrypi:~ $ <body>
-bash: syntax error near unexpected token `newline'
pi@raspberrypi:~ $ <p> list of databases:</p>
-bash: syntax error near unexpected token `newline'
pi@raspberrypi:~ $ <?php
-bash: ?php: No such file or directory
pi@raspberrypi:~ $ $link = mysql_connect('localhost', 'root', 'ruben');
-bash: syntax error near unexpected token `('
pi@raspberrypi:~ $ $res = mysql_query("SHOW DATABASES");
-bash: syntax error near unexpected token `('
pi@raspberrypi:~ $
pi@raspberrypi:~ $ while ($row = mysql_fetch_assoc($res)) {
-bash: syntax error near unexpected token `('
pi@raspberrypi:~ $     echo $row['Database'] . "<br>";
[Database] . <br>
pi@raspberrypi:~ $ }
-bash: syntax error near unexpected token `}'
pi@raspberrypi:~ $ ?>
-bash: syntax error near unexpected token `newline'
pi@raspberrypi:~ $ </body>
-bash: syntax error near unexpected token `newline'
pi@raspberrypi:~ $ </html>
-bash: syntax error near unexpected token `newline'
pi@raspberrypi:~ $ sudo nano /var/www/html/hellodb.php
```

The following screenshot has the script that has the code to access the database, connect to the server, and get the data from it:

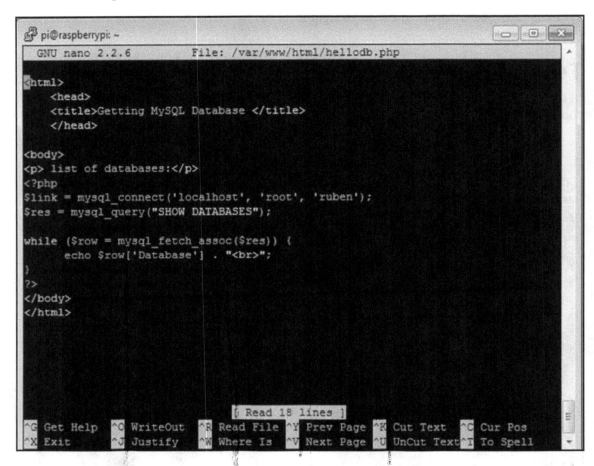

To test the page that we have made, type the IP address of your Raspberry Pi: `http://192.168.1.105/hellodb.php`. The page that appears is shown in the following screenshot, and you can now test that you can communicate between PHP and MySQL:

Installing phpMyAdmin for administrating databases

In this section, we will talk about how to configure your phpMyAdmin instance to administrate your database from a remote panel.

It's important that we install the client and the PHP5 module on the Apache server, so type the following command:

```
sudo apt-get install mysql-client php5-mysql
```

Next we will install the phpmyadmin package with the following command:

```
sudo apt install phpmyadmin
```

In the following screenshot, we can see the configuration of the server; in this case, we need to select apache2:

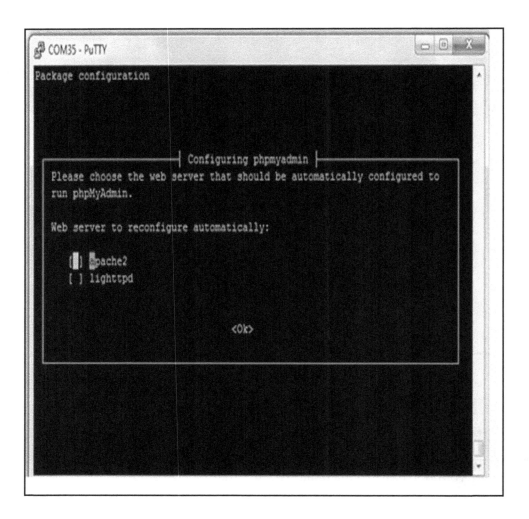

After that, we can select the database; in this case we select <No>:

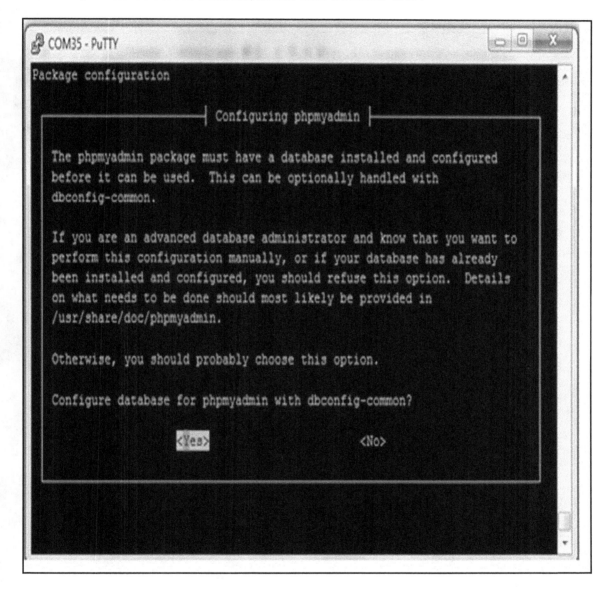

Configuring the Apache server

It's necessary that we configure the `apache2.conf` file. First, go to the Terminal on your Pi:

```
Sudo nano /etc/apache2/apache2.conf
```

In the following screen, we need to add the code:

```
Include /etc/phpmyadmin/apache.conf
```

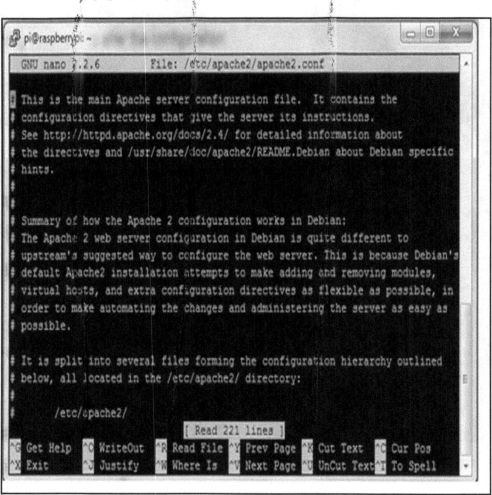

We include the following line at the bottom of the file:

```
Include /etc/phpmyadmin/apche.conf
```

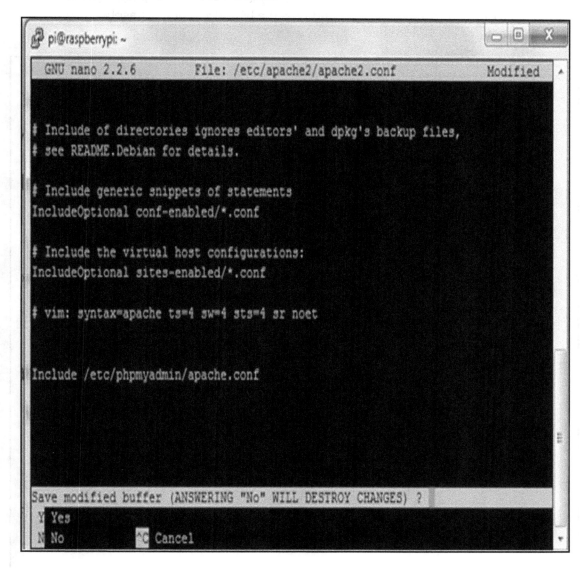

We have finally finished installing our Apache server, and we are now ready for the next step.

Entering the phpMyAdmin remote panel

After we have configured the server, we will enter the phpMyAdmin remote panel. We need to open our favorite web browser and type the IP address of our Raspberry Pi `http://Raspberry Pi Address/phpmyadmin`:

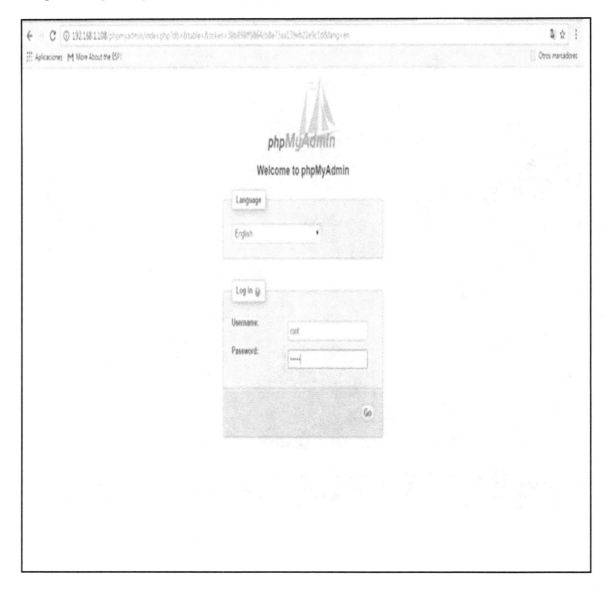

Integrating the Raspberry Pi and camera into the database

In this section, we'll see how to send data to the server and integrate the elements that we have seen during previous chapters. In the following image, we can see the finished integration of the system:

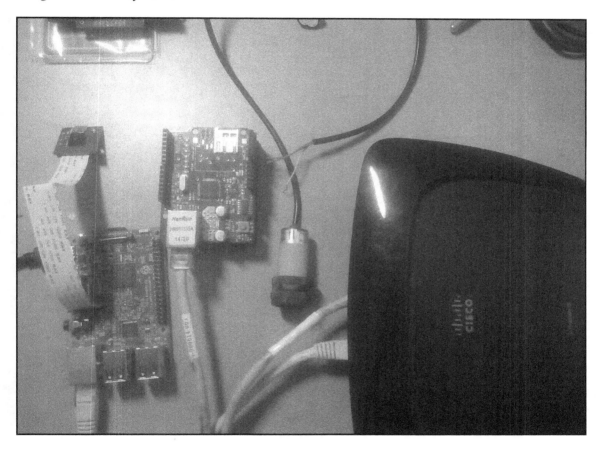

This is the Sketch for the Arduino board:

```
// Example web client using the Ethernet shield

// Include libraries
#include <SPI.h>
#include <Ethernet.h>
```

```
// Enter a MAC address for your controller below.
byte mac[] = { 0x90, 0xA2, 0xDA, 0x0E, 0xFE, 0x40 };

//IpAddress if DHCP fails
IPAddress ip(192,168,1,50);

// IP address of the Raspberry
IPAddress server(192,168,1,108);

// Initialize the Ethernet client
EthernetClient client;

void setup() {
 // Open serial communications
 Serial.begin(9600);

 // Start the Ethernet connection
 //if (Ethernet.begin(mac) == 0) {
 // Serial.println("La Dirección IP se asigna por DHCP");
 Ethernet.begin(mac, ip);

 // Display IP
 Serial.print("IP address: ");
 Serial.println(Ethernet.localIP());

 // Give the Ethernet shield a second to initialize
 delay(1000);
 Serial.println("Conectando...");
}
void loop()
{
 //variables to deploy
 Serial.println("Humidity: " + hum);

 if (client.connect(server, 80)) {
 if (client.connected()) {
 Serial.println("connected");

 // Make a HTTP request:
 client.println("GET /parking.php?valor1=" + valor1 + "&valor2=" +
valor2 + " HTTP/1.1");
 client.println("Host: 192.168.1.108");
 client.println("Connection: close");
 client.println();
 }
 else {
 // If you didn't get a connection to the server
 Serial.println("failed the connection");
```

```
}

// Read the answer
while (client.connected()) {
while (client.available()) {
char c = client.read();
Serial.print(c);
}
}

// If the server's disconnected, stop the client:
if (!client.connected()) {
Serial.println();
Serial.println("desconectado.");
client.stop();
}
}
// Repeat every second
delay(5000);
}
```

Programming the script software

In the following code, we have a script that will communicate with the Arduino board, and it is installed on the server. You can now either copy the following code inside a file called parking.php, or get the complete code file from the folder for this project:

```php
<?php
if (isset($_GET["model"]) && isset($_GET["plate"])) {
$temperature = intval($_GET["model"]);
$humidity = intval($_GET["plate"]);
$con=mysql_connect("localhost","root","ruben","arduinobd");
mysql_select_db('arduinobd',$con);
if(mysql_query("INSERT INTO data (model, plate)
VALUES ('$model', '$plate');")){
echo "Data were saved";
}
else {
echo "Fail the recorded data";
}
mysql_close($con);
}
?>
```

Node Express app for openALPR

If we want to send data using Node.js, we can use the following script, from `https://github.com/gerhardsletten/express-openalpr-server`:

```
POST /plates with json:
{
        image: (base64 encoded data),
        country_code: 'eu',
        pattern_code: 'no'
}
returns
{
        plate: 'DP49829',
        confidence: 79.758995,
        matches_template: 1,
        plate_index: 0,
        region: 'no',
        region_confidence: 0,
        processing_time_ms: 66.374001,
        requested_topn: 10,
        coordinates : {
                ...
        },
        candidates : [
                { plate: 'DP498Z9', confidence: 87.328056,
matches_template: 0 },
                { plate: 'DP98Z9', confidence: 83.397873, matches_template:
0 },
                ...
        ]
```

Amazon Web Services IoT

AWS IoT provides secure, bidirectional communication between internet-connected things (such as sensors, actuators, embedded devices, or smart appliances) and the AWS cloud. This enables you to collect telemetry data from multiple devices and store and analyze the data. You can also create applications that enable your users to control these devices from their phones or tablets.

AWS IoT components

In this section, we'll describe the components of AWS IoT:

- Device gateway
- Message broker
- Rules engine
- Security and identity service
- Thing registry
- Thing shadow
- Thing shadow service

Accessing AWS IoT

AWS IoT provide the following tools to create and communicate with things:

- **AWS command-line interface** (**AWS CLI**): Run commands for AWS IoT on Windows, macOS, and Linux. These commands allow you to create and manage things, certificates, rules, and policies. To get started, see the AWS CLI user guide (http://docs.aws.amazon.com/cli/latest/userguide/cli-chap-welcome.html). For more information about the commands for AWS IoT, see **iot** (http://docs.aws.amazon.com/cli/latest/reference/iot/index.html) in the **AWS Command Line Interface Reference**.
- **AWS IoT APIs**: Build your IoT applications using HTTP or HTTPS requests. These APIs allow you to programmatically create and manage things, certificates, rules, and policies. For more information about the API actions for AWS IoT, see **Actions** (http://docs.aws.amazon.com/iot/latest/apireference/API_Operations.html) in the AWS IoT **API Reference**.
- **AWS SDKs**: Build your IoT applications using language-specific APIs. These SDKs wrap the HTTP/HTTPS API and allow you to program in any of the supported languages. For more information, see **AWS SDKs and Tools** (https://aws.amazon.com/tools/#sdk).
- **AWS IoT Device SDKs**: Build applications that run on your devices that send messages to and receive messages from AWS IoT. For more information see, **AWS IoT SDKs** (http://docs.aws.amazon.com/iot/latest/developerguide/iot-sdks.html).

Services

AWS IoT integrates with the following services:

- **Amazon Simple Storage Service**: Provides scalable storage in the AWS cloud. For more information, see **Amazon S3** (`https://aws.amazon.com/s3/`).
- **Amazon DynamoDB**: Provides managed NoSQL databases. For more information, see **Amazon DynamoDB** (`https://aws.amazon.com/dynamodb/`).
- **Amazon Kinesis**: Enables real-time processing of streaming data at a massive scale. For more information, see **Amazon Kinesis** (`https://aws.amazon.com/kinesis/`).
- **AWS Lambda**: Runs your code on virtual servers from Amazon EC2 in response to events. For more information, see **AWS Lambda** (`https://aws.amazon.com/lambda/`).
- **Amazon Simple Notification Service**: Sends or receives notifications. For more information, see **Amazon SNS** (`https://aws.amazon.com/sns/`).
- **Amazon Simple Queue Service**: Stores data in a queue to be retrieved by applications. For more information, see **Amazon SQS** (`https://aws.amazon.com/sqs/`).

In the next section, we will use the Amazon **Simple Notification Service** (SNS) in our project.

Setting up the Raspberry Pi with AWS IoT

Follow these steps to connect your Raspberry Pi to the AWS IoT platform. It's recommended you go to `http://docs.aws.amazon.com/iot/latest/developerguide/iot-sdk-setup.html`.

AWS IoT connections

Now it's time to connect your Raspberry Pi 3 to the IoT service. To do that, you need to login to AWS, select **IoT**, and start a wizard by configuring a device:

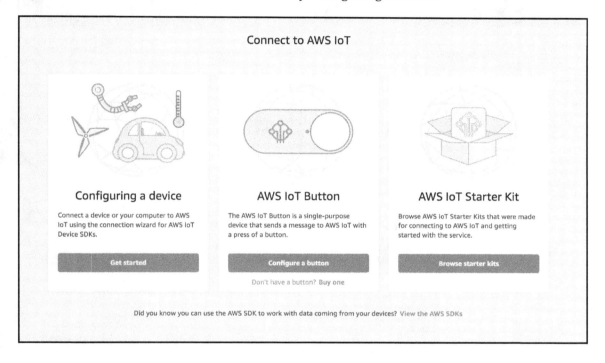

Select the **Get started** option as illustrated in the following screenshot:

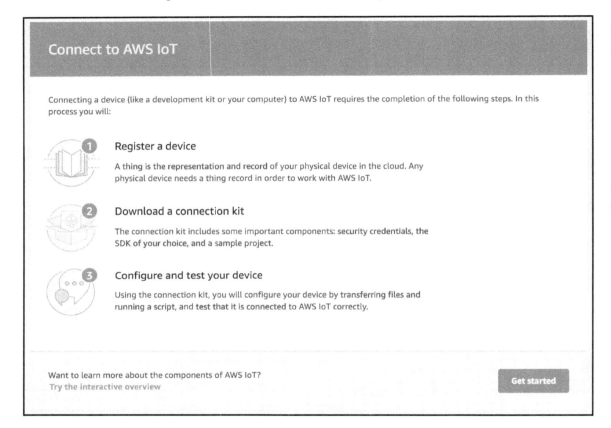

Select the platform to be connected (**Linux/OSX** | **Python**):

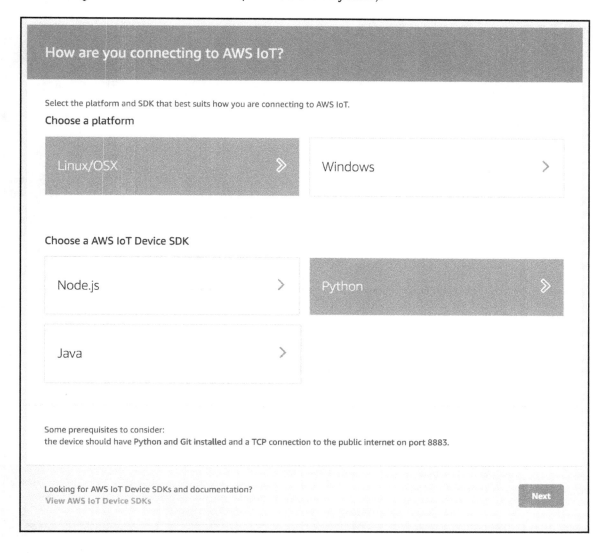

Register the thing:

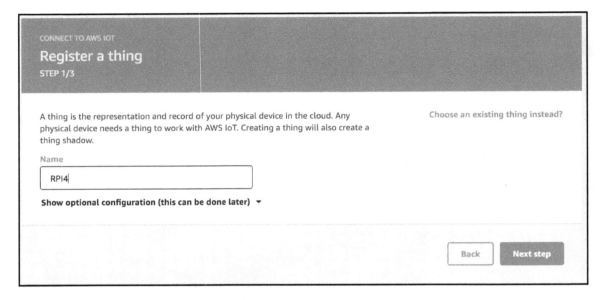

Next, we **Download a connection kit**:

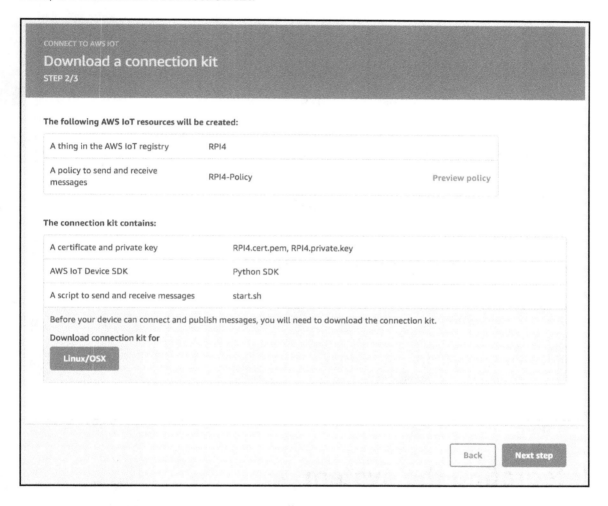

The final step is you click on **Done**:

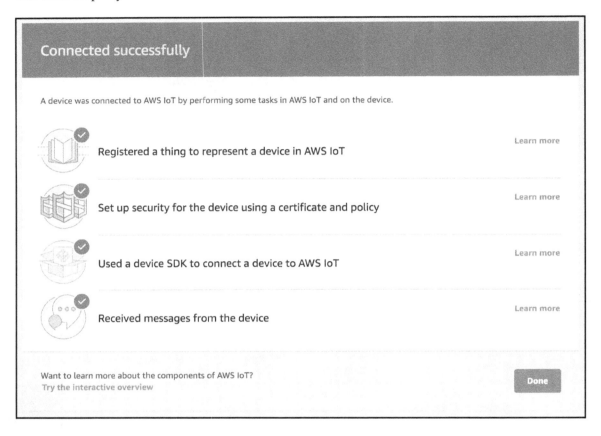

Integrating the system

To integrate the system, we will have a schema to see how the system could be implemented using the information and requirements seen in the previous section. In the following figure, we see the modules or steps that we need to follow to build the complete system of the parking lot:

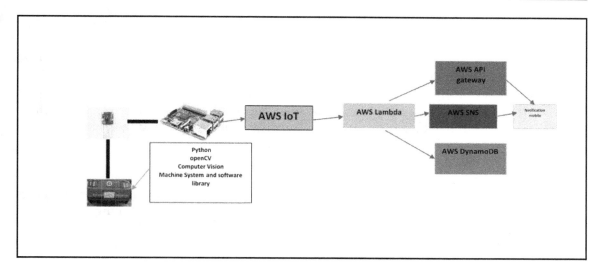

1. The camera detects the plate number using the library software OpenCV and Python.
2. The Raspberry Pi is connected to Amazon Web Services and we configure the Python SDK for receiving commands; it is connected to the IoT Lambda services that receives the commands.
3. From a Python script, the AWS IoT service receives the data captured from the camera and is stored in AWS DynamoDB, which is a database service. It stores the ID of the car and the plate number of the vehicle.
4. The data captured from the vehicle is stored in the web service; it will count the number of cars that go in or out of the parking area, and also counts the total number of places in the parking lot.
5. We create a trigger that will be executed when places are vacant, according to the number calculated in the previous step.
6. The notification system will send an SMS when a vacant place is created.

Future parking systems

For ideas that will be applied to parking lot systems in the future, consider some important additional features that will be offered to make parking easier for drivers:

- **Payment modes**: Both on-street and off-street space charge. This entire transaction can be made from a mobile phone. Have a system where the charge is paid in intervals of time, and if it crosses the limit it will send an SMS alert to the towing company/traffic police for ticketing or towing.

- **Parking reservation**: Instead of driving to the lot and then getting directions, the customers will be able to reserve parking from their home, even before leaving. For this, the server will need an Internet connection and must be able to recognize the appropriate customer.
- **Connected lighting**: With the use of sensors, parking lots are kept illuminated throughout the day. With an automated system, this will change. The places that are vacant will be dimmed to reduce energy consumption and the lights will be off when the lot is empty; the system will detect when a car is present and turn on the lights.
- **GPS-based directions**: With the use of GPS systems, drivers can go and find a specific parking lot. The user will be able to get real-time directions and guidance to the space. On a mobile phone, they will see the map of the parking lots that best fits each car and can park in a specific place.
- **Better governance and traffic jam avoidance**: This will apply to parking lots that can monitor people, using security systems.
- **For local government bodies**: Patterns of high and low parking density could result in varying tariff rates to discourage customers from using cars on certain days, thereby reducing traffic congestion. Variable tariff rates could also encourage car pooling depending on the time of day.

Summary

In this chapter we learned how to build a smart parking system using some hardware and software stuff like sensors, camera, electronics, and a Raspberry Pi as a central interface, and we used Arduino to detect when a car enters the parking. After that, we built a parking system plan with some elements for detecting and logging data in a database, which can help us know the number of cars that are in the and how many parking spaces are left; at the end, we mentioned some steps to follow in order to integrate the system with AWS IoT services; for example, the project can be improved by detecting vacant places using a camera and determining which specific places are vacant by taking a picture of the entire parking lot.

In the next chapter, we will see how to build a home system: Netflix's The Switch for the living room. It will be an interesting project!

5
Building Netflix's The Switch for the Living Room

On the one day off that we want to rest and be relaxed at home, everything in our house needs to be controlled without standing up or moving something else with our hands. In this case, we are in our living room or any place at home and everything is going on and moving manually, but sometimes, people like to be comfortable, for example, turning on the lights, watering the plants in the garden, opening the garage door, if a person is in front of the door house, we want to know who is ringing the doorbell, send or calling for a service, or sending an SMS to open the door; just with one click, we can do many actions, save our lives, help people have a better mood.

In this chapter, we will build a *smart button* that can do things automatically for a home automation system. This can be used at home to control our electrical devices, order some services from outside, and many more things.

In this chapter, we will cover the following topics:

- Setting up the Particle Photon (Wi-Fi)
- Hardware and software requirements
- The architecture of the home automation system
- Reading IR signals
- Assembling the electronics
- Controlling smart lights
- Sending notifications with IFTTT
- Future ideas for projects

Setting up the Particle Photon

In this section, we will the board Particle Photon, that is the base and is the main brain of this interesting project. It's based on the Arduino IDE programming software and in the following sections we will learn the architecture and the main parts to configure it for our real application. One of the most important features it has, is the connectivity with Wi-Fi. This is one of the most important points to consider, so we can configure the board as a Wi-Fi Device. We will see more details next.

Getting started

To ready our board, we need to know some important notes about the Particle Photon.

The main brain is a microcontroller and it's inside the board. It works as an Arduino board, it has firmware inside the board and we make a software and we download an embedded application. We connect the hardware using some cables and interact with the hardware with the software programming code and the pins interact with the outside to move a real application. Before we continue with the next part, we first need to see some important things about Particle Photon in order to build our project without problems.

What's in the box?

The kit of the Particle Photon has the following parts, and we can get it at `https://store. particle.io/`:

- The Wi-Fi module
- You need some LEDS
- Resistors
- Photoresistor
- Push buttons
- Cables
- Protoboard

The following figure shows the kit of the Particle Photon:

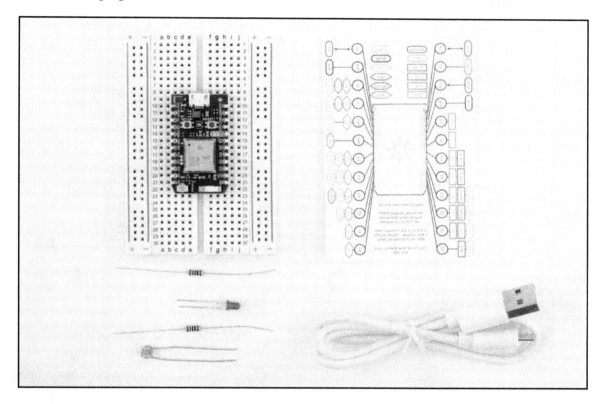

The architecture of the Particle Photon

In this section will see the hardware of the Particle Photon, this means the pins for the inputs or output connected to the board. We can connect to the input pins, push buttons, or sensors and they will be listed to the real world signals. By the output they mean the connected devices that will activate when something happens, for example, a buzzer that can see be on when something happens. In the following image we can the pins of the hardware architecture of the board and the names of the pins that it has:

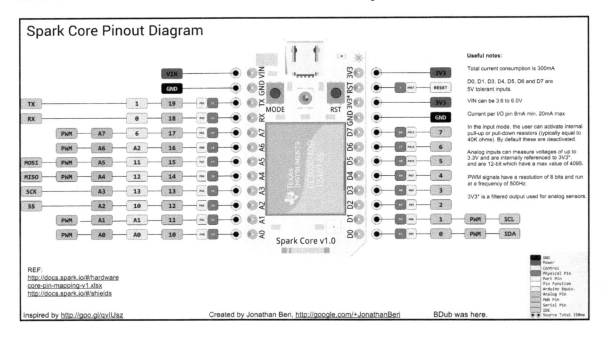

Requirements

Now we have the most important requirements that we need to ready our Particle Photon board.

Software

Download the Spark Core mobile app:

- **iPhone**: https://itunes.apple.com/us/app/spark-core/id760157884
- **Android**: https://play.google.com/store/apps/details?id=io.spark.core.android
- **Windows**: https://www.microsoft.com/en-us/store/p/particle/9nblggh4p55n

We highly recommend that you use the mobile app for the first-time setup.

Hardware

In this part, we will mention how to prepare our Particle Photon in order to get it ready to work and talk about the stuff that the box has:

- Your Particle device, brand new and out of the box
- USB to micro USB cable
- Power source for USB cable (such as your computer, USB battery, or power brick)
- Your iPhone or Android or Windows smartphone

Wi-Fi settings

We will see the Wi-Fi connection settings of the board:

- 2.4 GHz capable router
- Select one of the channels from 1-13
- WPA/WPA2 encryption
- On a broadcast SSID network
- Not behind a hard firewall or enterprise network

We do not recommend using WEP Wi-Fi settings for security reasons.

Specifications of the board

Building your own board for a specific Internet of Things project has many advantages. Well, first, it means that you don't end up with an ugly breadboard with wires going everywhere. This also adds a layer of safety and reliability to your project, as no wire can get disconnected during the use of the project. Finally, it means that if your board is also interesting for other people, you already have a project that's nearly ready for production and reselling.

The board itself has several digital and analog I/Os as well as an onboard Wi-Fi chip and processor. But that's not all: by default, the Photon board connects automatically to the Particle platform, which is an online cloud platform from which you can control your board. Therefore, you don't have to find a cloud platform provider for the project; it's all done for you automatically.

Now, an important thing at this point is to make sure your Photon is updated to the latest version. The one I received had an old firmware on it, and I just couldn't program it from the web IDE. To do that, I recommend that you use the Particle CLI tool. Install it from `https://github.com/spark/particle-cli`.

You also need to put your Photon in the DFU mode. To do that, simply hold both buttons on the board and then release the *reset* button first. The Photon LED should be blinking yellow.

Finally, download the firmware and get the required instructions from `https://github.com/spark/firmware/releases`.

In this project, we are going to use the Particle Photon board itself but in the version that comes without the headers. With this version, we'll be able to integrate it into a PCB design.

Connecting your Particle Photon to the internet using the setup web application

To continue, we need to follow these steps:

 It's important to have a an account user, you need to create one on the particle web site: https://www.particle.io/

1. Go to the URL https://www.setup.particle.io/.
2. Click on **Setup a Photon**.
3. Connect your PC to the Particle Photon by connecting to the network name PHOTON.
4. Open the file that you want to test.
5. Configure your Wi-Fi connections.
6. Rename your device.

 This process works only in Chrome/Firefox/Opera.

If you want to connect your Particle Photon from a mobile phone, you can go to https://docs.particle.io/guide/getting-started/start/core/.

Blinking an LED

Connect everything, as shown in the following diagram:

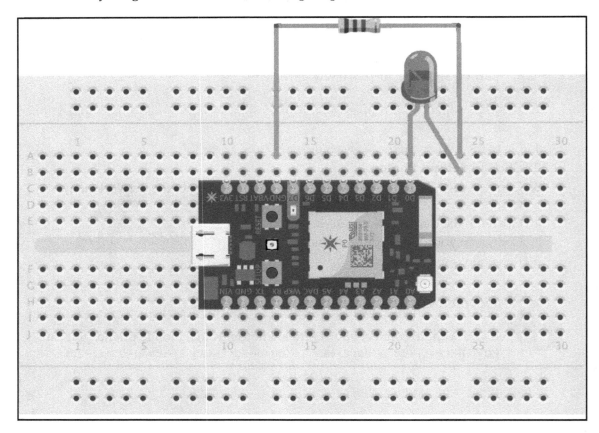

Hardware

We need the following hardware:

- Your Particle device
- USB to micro USB cable
- Power source for USB cable (such as your computer, USB battery, or power brick)

- Two resistors between 220 Ohms and 1,000 Ohms
- One LED, any color
- One photoresistor

Software

Also, we need the following software:

- The online IDE: `http://build.particle.io/`
- Local Particle Dev: `http://particle.io/dev`

Testing the code

To test the code, we need to add a new device and then a flash on the board:

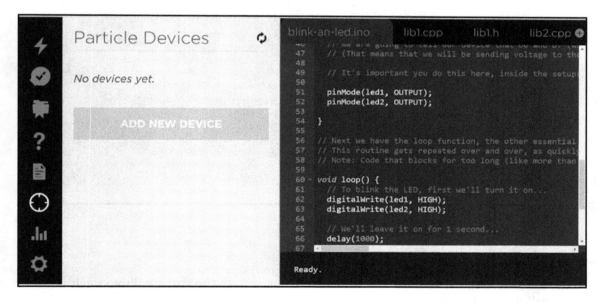

If everything is connected, the LED connected to the board will have to turn on. After all of this, now we are ready and can move to the next step. We can go to the next section and start building our project.

Hardware and software requirements

To build this project, we need the following materials:

- Particle Photon
- Momentary Arcade button
- 5 mm IR LED 940 nm
- Transistor NPN (BC547)
- 3.7 volts 1000 mAh LiPo battery
- LiPo battery charger
- 4.7K Ohm resistor
- Philips Hue lamp with a wireless bridge

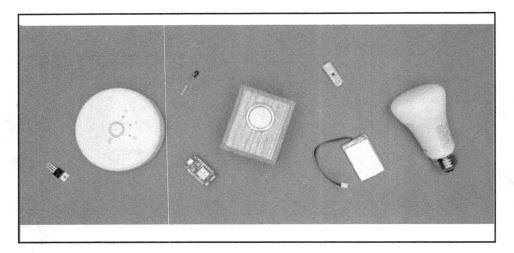

Hardware configuration

In the following figure, we have the connections made with the Particle Photon, with the push button, and the IR infrared LED connected with the transistor:

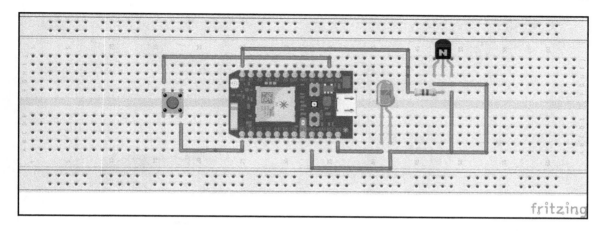

The architecture of the home automation system

In the following figure, we show the elements that integrate the home automation system:

- First, we have the Particle Photon that sends the signal when the button is pressed via the infrarred sensor connected to the board.
- The TV sensor (**IR receiver sensor**) receives the coded signal and turns on the TV and enables the Netflix system
- All the systems are connected to a home Wi-Fi network (Particle Photon and bulbs are also connected to the network)
- Also, when the button is pressed, the lights will be on or off
- The control of the system is based on the Particle Photon through the infrared sensor when the button is pressed

- All the devices connected to the system can be controlled from the main board, and the main software is created according to the devices and services we want to execute or send data

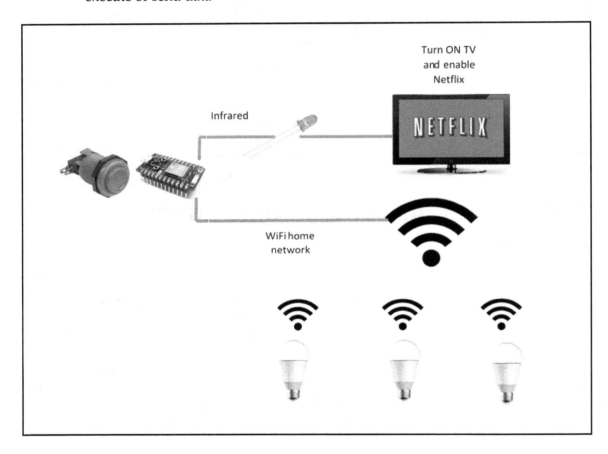

Reading IR signals

In this section, we will explain how code signals are sent to the TV; the following code is made for a TV recommended for Netflix that has a button to enable Netflix; in this case, you can use any remote control. In this project, we will use a Philips TV in order to do that, we can start turning on your TV and start Netflix in one swift motion, you'll need to send the same IR signal as the Netflix button on your TV remote. If you're not using a TV with a Netflix button or if you have an older TV, this may involve pressing more than one button, and you'll need to adjust accordingly.

We will get the IR signals to read them. We used an IR receiver and an Arduino.

Sensor IR receiver

We need to test our remote control to get code signals in order to get proper code to communicate with our TC. In this case, we will use a Philips remote control TV; we connect our sensor TSOP382 IR receiver diode from Vishay (http://www.vishay.com/), as shown in the following diagram:

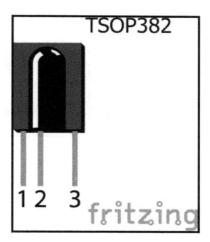

Hardware connections

Connect the sensor IR receiver to the Arduino board, as shown in the following figure:

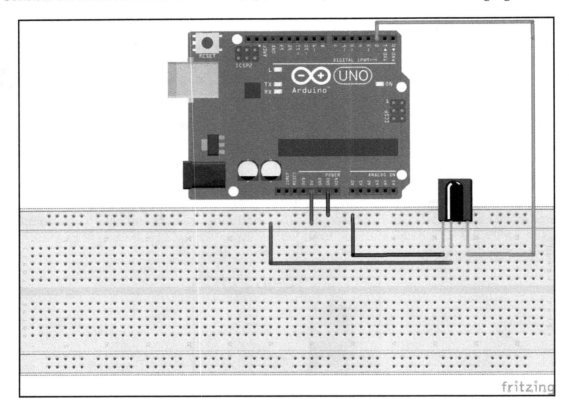

Code to read IR signals

After this, open the Arduino IDE and you will have the sketch that will help us decode the signal for the Netflix key in our remote control:

```
/* Raw IR decoder sketch!
This sketch/program uses the Arduino and a PNA4602 to decode IR received.
This can be used to make a IR receiver (by looking for a particular code)
or transmitter (by pulsing an IR LED at ~38KHz for the durations detected

Code is public domain, check out www.ladyada.net and adafruit.com for more
tutorials!
*/
```

```
// We need to use the 'raw' pin reading methods
// because timing is very important here and the digitalRead()
// procedure is slower!
//uint8_t IRpin = 2;
// Digital pin #2 is the same as Pin D2 see
// http://arduino.cc/en/Hacking/PinMapping168 for the 'raw' pin mapping

#define IRpin_PIN PIND
#define IRpin 2

// for MEGA use these!
//#define IRpin_PIN PINE
//#define IRpin 4

// the maximum pulse we'll listen for - 65 milliseconds is a long time
#define MAXPULSE 65000

// what our timing resolution should be, larger is better // as its more
'precise' - but too large and you wont get // accurate timing

#define RESOLUTION 20

// we will store up to 100 pulse pairs (this is -a lot-)
uint16_t pulses[100][2]; // pair is high and low pulse
uint8_t currentpulse = 0; // index for pulses we're storing

void setup(void) {
  Serial.begin(9600);
  Serial.println("Ready to decode IR!");
}

void loop(void) {
  uint16_t highpulse, lowpulse; // temporary storage timing
  highpulse = lowpulse = 0; // start out with no pulse length
// while (digitalRead(IRpin)) { // this is too slow!
    while (IRpin_PIN & (1 << IRpin)) {

      // pin is still HIGH
      // count off another few microseconds

      highpulse++;
      delayMicroseconds(RESOLUTION);

      // If the pulse is too long, we 'timed out' - either nothing
      // was received or the code is finished, so print what
      // we've grabbed so far, and then reset
       if ((highpulse >= MAXPULSE) && (currentpulse != 0)) {
        printpulses();
```

```
        currentpulse=0;
        return;
      }
  }
  // we didn't time out so lets stash the reading
  pulses[currentpulse][0] = highpulse;
  // same as above
  while (! (IRpin_PIN & _BV(IRpin))) {
     // pin is still LOW
     lowpulse++;
     delayMicroseconds(RESOLUTION);
     if ((lowpulse >= MAXPULSE) && (currentpulse != 0)) {
       printpulses();
       currentpulse=0;
       return;
     }
  }
  pulses[currentpulse][1] = lowpulse;

  // we read one high-low pulse successfully, continue!
  currentpulse++;
}

//Function to display on monitor serial

void printpulses(void) {
  Serial.println("\n\r\n\rReceived: \n\rOFF \tON");
  for (uint8_t i = 0; i < currentpulse; i++) {
    Serial.print(pulses[i][0] * RESOLUTION, DEC);
    Serial.print(" usec, ");
    Serial.print(pulses[i][1] * RESOLUTION, DEC);
    Serial.println(" usec");
  }
  // print it in a 'array' format
  Serial.println("int IRsignal[] = {");
  Serial.println("// ON, OFF (in 10's of microseconds)");
  for (uint8_t i = 0; i < currentpulse-1; i++) {
    Serial.print("\t"); // tab
    Serial.print(pulses[i][1] * RESOLUTION / 10, DEC);
    Serial.print(", ");
    Serial.print(pulses[i+1][0] * RESOLUTION / 10, DEC);
    Serial.println(",");
  }
  Serial.print("\t"); // tab
  Serial.print(pulses[currentpulse-1][1] * RESOLUTION / 10, DEC);
  Serial.print(", 0};");
}
```

After you download the sketch to your Arduino board, place your remote control in front of your IR receiver sensor. Press the *ON* key and this is the result that follows:

```
340 usec, 520 usec
340 usec, 520 usec
340 usec, 520 usec
340 usec, 500 usec
360 usec, 500 usec
340 usec, 520 usec
340 usec, 520 usec
340 usec, 940 usec
340 usec, 520 usec
760 usec, 520 usec
340 usec, 520 usec
int IRsignal[] = {
// ON, OFF (in 10's of microseconds)
        274, 80,
        52, 76,
        52, 34,
        52, 34,
        138, 120,
        52, 34,
        52, 34,
        50, 34,
        52, 34,
        52, 34,
        52, 34,
```

If we want to decode the Netflix key in this, note that I used a Philips TV and this is the result:

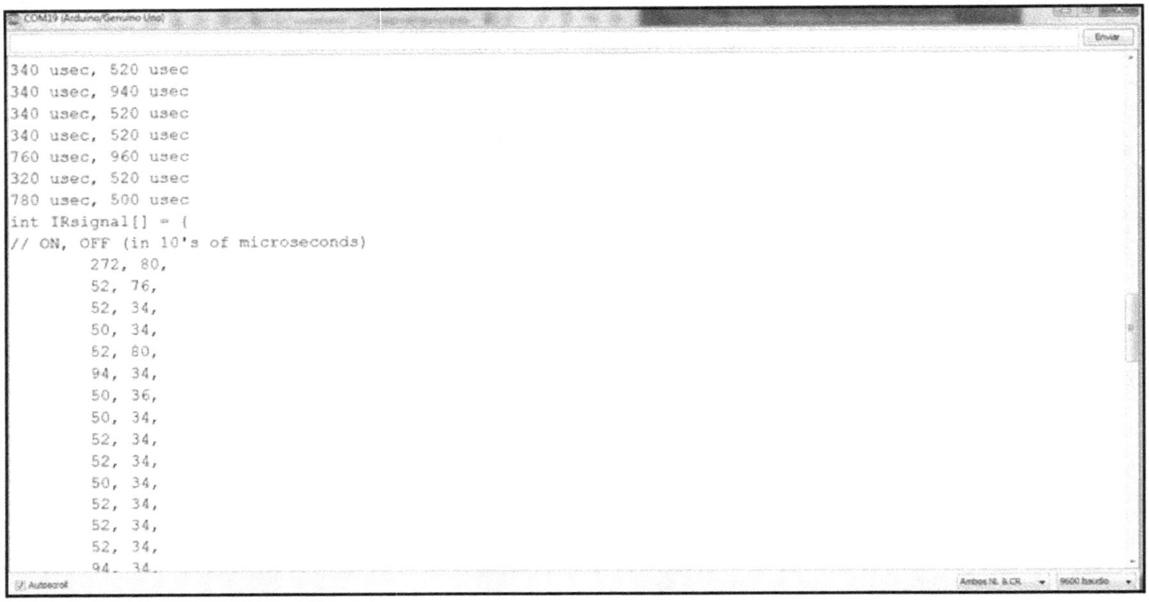

Programming the button

We're are programming a Netflix TV remote control, and we see the code for testing the sending signals from the Particle Photon to the TV.

Testing the code for remote control for a Netflix TV

In the following sketch, we show an example of programming the button:

```
//Before using, make sure you get your IR timings and add your hue bridge
IP/dev username in the appropriate places.

//To keep it as simple as possible, this does not contain any code for the
optional LEDs on the case. We light up our indicators and logo after the
button is pressed.
```

```
TCPClient client;

#define PIN_IR A0
#define PIN D0
#define NUM_PULSES 68

//specific to your TV
int pulse_widths[NUM_PULSES][2] = {
 {62116, 8860}, {4360, 540}, {520, 540}, {1620, 560}, {500, 560}, {1600,
560},    {500, 560},{1600, 560}, {1600, 560}, {1640, 520}, {1640, 520},
{1640, 540},
{1620, 520},{540, 540}, {500, 560}, {500, 560}, {1640, 520}, {1600, 560},
{520, 540},{1620, 540}, {540, 540}, {500, 560}, {1600, 560}, {500, 560},
{1620, 540},
 {520, 540}, {1620, 560}, {520, 520}, {1620, 560}, {1600, 560}, {540, 520},
 {1600, 560}, {520, 540}, {1600, 560}, {37200, 8860}, {4340, 560}, {540,
520},
 {1600, 560}, {520, 560}, {1600, 560}, {500, 560}, {1600, 560}, {1600,
560},
 {1640, 520}, {1640, 520}, {1640, 520}, {1640, 540}, {500, 540}, {540,
520},
 {520, 560}, {1600, 560}, {1620, 540}, {520, 540}, {1620, 540}, {520, 560},
 {500, 560}, {1640, 520}, {500, 560}, {1620, 540}, {1620, 540}, {1620,
560},
 {520, 540}, {1600, 560}, {1600, 540}, {560, 520}, {1640, 520}, {540, 540},
 {500, 540}
};

//hue stuff
//do this first
http://www.developers.meethue.com/documentation/getting-started
//make this the IP of your bridge

byte hueBridge[] = {10,0,1,176};
char hueString[100];

//state
int reading = 0;

void setup() {
    pinMode(PIN, INPUT_PULLUP);
    pinMode(PIN_IR, OUTPUT);
}
void loop() {
    //This could also be an interrupt
    reading = digitalRead(PIN);
    if(reading == 0){
        IR_transmit_pwr();
```

```
        delay(2000);

//publish for any listening applications (for advanced stuff / extensions)
        Spark.publish("switch:trigger", "ON", 0, PRIVATE);
        doLight(false);
        delay(5000);
    }else{
        delay(100);
    }
    // RGB.control(false);
}
void doLight(bool turnOn){
    if(!client.connected()){
        client.connect(hueBridge, 80);
        if(client.connected()){
            if(turnOn){

  //turn light on. Can pass HSV or delay time as well in the payload.
    String payload = "{\"on\":true}";
    sprintf(hueString, "PUT /api/YOUR_DEV_USERNAME/groups/0/action
HTTP/1.1\r\nContent-Length:%i\r\nContent-
Type:application/json\r\n\r\n%s\r\n", payload.length(), payload.c_str());
            client.print(hueString);
        }else{

  //turn light off. Can pass HSV or delay time as well in the payload.
  String payload = "{\"on\":false}";
  sprintf(hueString, "PUT /api/YOUR_DEV_USERNAME/groups/0/action
HTTP/1.1\r\nContent-Length:%i\r\nContent-
Type:application/json\r\n\r\n%s\r\n", payload.length(), payload.c_str());
            client.print(hueString);
        }
        client.flush();
        client.stop();
      }
    }
}

//based on src from https://learn.adafruit.com/ir-sensor
void pulseIR(long microsecs) {
  while (microsecs > 0) {
    // 38 kHz is about 13 microseconds high and 13 microseconds low
    digitalWrite(PIN_IR, HIGH); // this takes about 3 microseconds to happen
    delayMicroseconds(10); // hang out for 10 microseconds
    digitalWrite(PIN_IR, LOW); // this also takes about 3 microseconds
    delayMicroseconds(10); // hang out for 10 microseconds
    // so 26 microseconds altogether
    microsecs -= 26;
```

```
    }
}

void IR_transmit_pwr() {
    for (int i = 0; i < NUM_PULSES; i++) {
        delayMicroseconds(pulse_widths[i][0]);
        pulseIR(pulse_widths[i][1]);
    }
}
```

Testing the code for a Philips TV

Now, after we create the test to get the code for a specific remote control, here we have the code showed in an array, and this is the result of that action:

```
int IRsignal[] = {// ON, OFF (in 10's of microseconds)
272, 80, 52, 76,52, 34,50, 34,52, 80,94, 34,50, 36, 50, 34,52, 34,52,
34,50, 34, 52, 34,52, 34,52, 34,94, 34,52, 32,52, 78, 94, 34,52, 78,50,
1560,274, 80, 52, 76,52, 34,52, 34,48, 82,94,34,52, 34,52, 34,50, 34,52,
34,52, 34,52, 34,50, 34,52, 34,94, 34,52, 34,52, 76, 96, 34,50, 78,52,
1558,276, 80,50, 78,52, 34,50, 34,52, 80,94, 32,52, 34,52, 34,52, 34,50,
36,50, 34,52, 34,52, 34,          52,32,96, 34,50, 34,52, 78,94, 34,52,
76,52, 1560,274, 80,52, 76,52, 34,52, 34,50, 80,94,34,52, 34, 52, 32,52,
34,52, 34,52, 34,50, 36,50, 34,52, 34,94, 34,52, 34,52, 76,96, 32,52,
78,50, 0};
```

This code needs to be added in the sketch in the array where it gets the data; this will be sent by the Photon board to the TV via the infrared sensor to the IR receiver sensor of the TV.

Assembling the electronics

Now we will look at the hardware part; this is to help us build our circuit. First, we need to do this in a breadboard and then solder the materials.

The circuit to be built

This the final circuit of the project, and it can be used to put it; first in a breadboard and then to be used as a prototype:

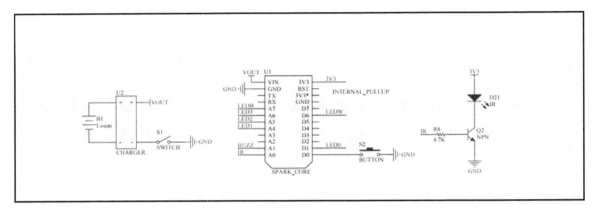

The final circuit

In the following figure, we can see the circuit that can be soldered according to the project:

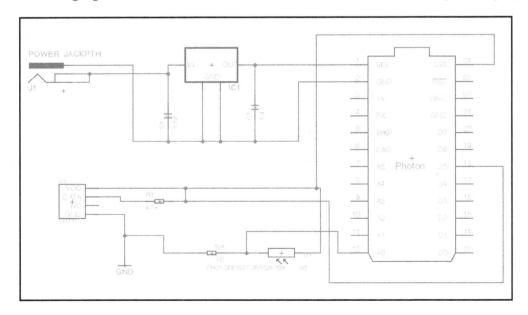

The circuit layout

In the following figure, we see the circuit layout:

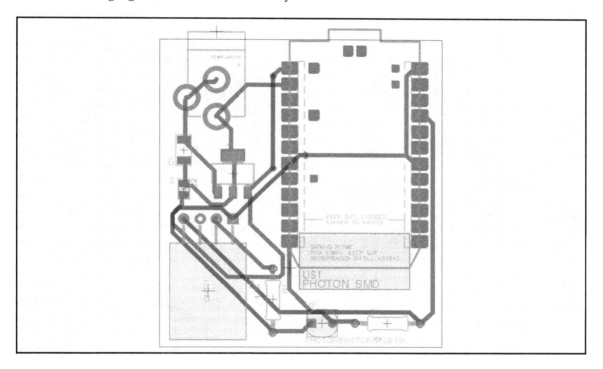

Controlling smart lights

In this section, we will describe the important steps to follow in order to configure the devices to control the lights with the button.

Getting started

In the following steps, we will look at how to configure our bridge device to receive commands and set up the lights to control them.

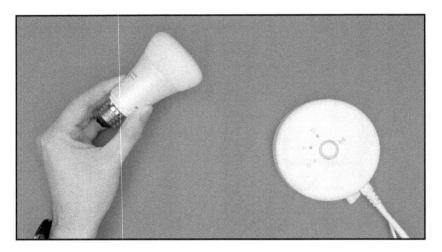

In the following sections, we will look at the steps to configure the bridge device and the API for the connected devices in order to connect them to the network; this information can be found at `https://developers.meethue.com/documentation`.

Step 1

First we need to connect and configure the bridge device, and the first thing we need to do is verify that it is connected.

Step 2

Then, you need to discover the IP address of the bridge on your network. You can do this in a few ways:

1. Use a UPnP discovery app to find Philips Hue in your network.
2. Use our broker server discover process by visiting (`http://www2.meethue.com/en-us/about-hue`).

3. Log in to your wireless router and look up Philips Hue in the DHCP table.

4. Follow the Hue app method. Download the official Philips Hue app. Connect your phone to the network the Hue bridge is on. Start the Hue app (the iOS described here). Connect the bridge to your WiFi network with an Ethernet cable. Use the app to find the bridge and try to control the lights. If all is working, go to the settings menu in the app. Go to **My Bridge**. Then, go to **Network** settings. Switch off the DHCP toggle. The IP address of the bridge will show. Note the IP address and then switch DHCP back on:

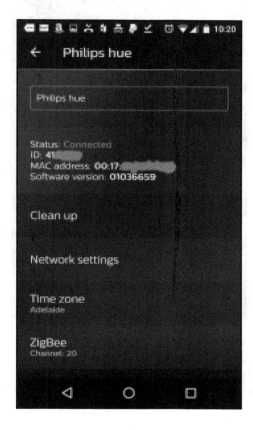

This is the IP address of the bridge and the DHCP:

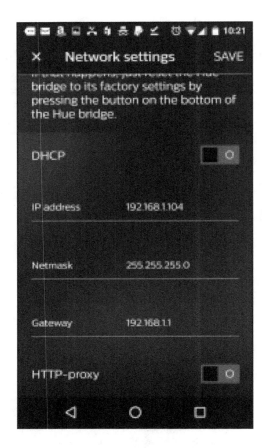

Step 3

Once you have the address, load the test app by visiting the following address in your web browser:

```
http://<bridge ip address>/debug/clip.html
```

You should see an interface like this:

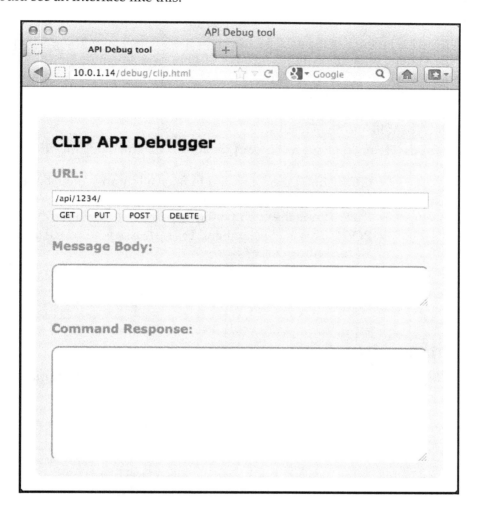

Using this debugger utility, you can populate the components of an HTTP call, the basis of all web traffic and the Hue RESTful interface:

- **URL**: This is actually the local address of a specific resource (thing) inside the Hue system. It could be light, a group of lights, or many more things. This is the object you'll be interacting with in this command.
- **Message body**: This is the part of the message that describes what you want to change and how. Here, you enter, in the JSON format, the resource name and the value you'd like to change/add.
- **Methods**: Here, you have a choice of the four HTTP methods the Hue call can use:
  - **GET**: This is the command to fetch all information about the addressed resource
  - **PUT**: This is the command to modify an addressed resource
  - **POST**: This is the command to create a new resource inside the addressed resource
  - **DELETE**: This is the command to delete the addressed resource
- **Command Response**: In this area, you'll see the response to your command.

The Philips Hue API

First, let's perform a very simple command and get information about your Hue system. Fill in the details, leaving the body box empty, and press the **GET** button:

Address	`http://<bridge ip address>/api/newdeveloper`
Body	
Method	`GET`

You should see a response similar to what is shown in the following screenshot:

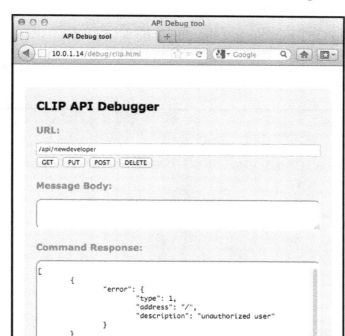

Now, this is the command to fetch all information in the bridge. You didn't get much back, and that's because you're using an unauthorized username: newdeveloper.

We need to use the randomly generated username that the bridge creates for you. Fill in the following information and press the **POST** button:

Address	http://<bridge ip address>/api
Body	{"devicetype":"my_hue_app#iphone peter"}
Method	POST

This command is basically asking you to create a new resource inside /api (where usernames sit) with the following properties.

When you press the **POST** button, you should get back an error message that lets you know that you have to press the link button in the application in order to test the device. This can help us to test the request and the app will control your lights. By pressing the button, we test whether the user has physical access to the bridge:

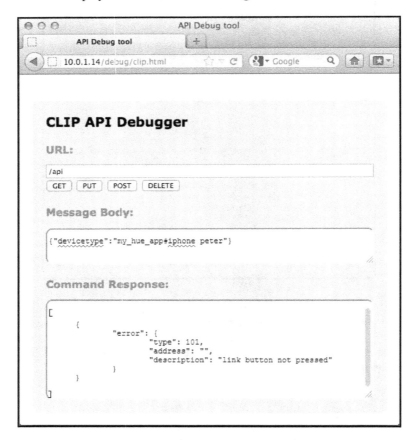

Go and press the button on the bridge and then press the **POST** button again and you should get a success response, as follows:

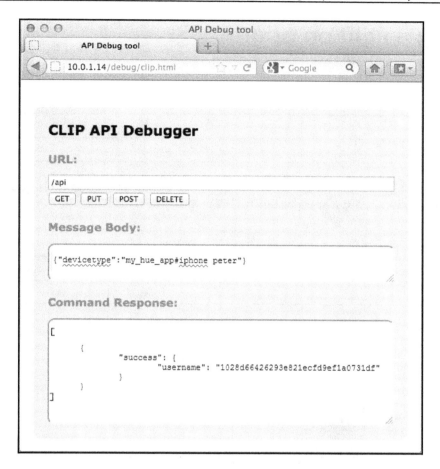

Congratulations, you've just created an authorized user
(1028d66426293e821ecfd9ef1a0731df), which we'll use from now on! Now if you
perform the first GET command again, you should get a whole lot more information about
what lights you have and their states. This data is all in the JSON format, so it can be easily
processed by your applications.

Turning a light on or off

Each light has its own URL. You can see what lights you have with the following command:

Address	`http://<bridge ip address>/api/1028d66426293e821ecfd9ef1a0731df/lights`
Body	
Method	GET

You should get a JSON response with all the lights in your system and their names. Now let's get information about a specific light. This is the light with ID 1:

Address	`http://<bridge ip address>/api/1028d66426293e821ecfd9ef1a0731df/lights/1`
Body	
Method	GET

In this response, you can see all of the resources this light has. The most interesting ones are inside the `state` object, as these are the ones we'll have to interact with in order to control the light.

Lets' start with the `on` attribute. This is a very simple attribute that can have two values: `true` and `false`. So let's try turning the light off:

Address	`http://<bridge ip address>/api/1028d66426293e821ecfd9ef1a0731df/lights/1/state`
Body	`{"on":false}`
Method	PUT

Looking at the command you are sending, you can see that we're addressing the `state` object of light one and asking it to modify the `on` value inside it to `false` (or `off`). When you press the **PUT** button, the light should turn off. Change the value in the body to `true`, and the light will turn on again.

Changing the intensity of the lights

Now let's do something a bit more fun and start changing some colors. Enter the following command:

Address	`http://<bridge ip address>/api/1028d66426293e821ecfd9ef1a0731df/lights/1/state`
Body	`{"on":true, "sat":254, "bri":254,"hue":10000}`
Method	`PUT`

We're interacting with the same `state` attributes here, but now we're modifying a couple more attributes. We're making sure the light is on by setting the `on` resource to `true`. We're also making sure the saturation (intensity) of the colors and the brightness are at their maximum by setting the `sat` and `bri` resources to `254`.

Finally, we're asking the system to set the `hue` (a measure of color) to `10000` points (Hue runs from 0 to 65,535). Try changing the hue value and keep pressing the **PUT** button and see the color of your light changing and running through different colors.

Sending notifications with IFTTT

The Particle channel on IFTTT will let you connect your devices to other powerful channels. You can now easily send and receive tweets, SMS, check the weather, respond to price changes, monitor astronauts, and much, much more. This page is a reference for you to use as you get your **Particle Recipes** set up.

Parts of an IFTTT

The following are the parts of IFTTT:

- **Triggers**: Trigger can be as conceptually simple as *"Is x greater than 5?"* or *"Did I get the new e-mail?"*
- **Actions**: Actions are what IFTTT does when the answer to your trigger question is yes! When set up, you can have IFTTT email you, post for you, save information to Dropbox, and many other useful functions.
- **Recipes**: Recipes are a combination of a **trigger** and an **action**. IFTTT lets you connect triggers to actions. Did someone tweet something interesting (that's a trigger) and then turn my disco ball on (that's an action)?

- **Ingredients**: Ingredients are pieces of data from a trigger. Ingredient values are automatically found by IFTTT using certain aspects of your device and/or firmware. These pieces of data can be used when setting up the action that goes with your created trigger. For Particle, ingredients will often include the name you've given to your Core or Photon, the time that the trigger occurred, and any data that the trigger returned.

Other IFTTT channels will provide (and sometimes automatically insert) their own ingredients. If ingredients are available, they can be found in the blue Erlenmeyer flask icon next to the relevant input box:

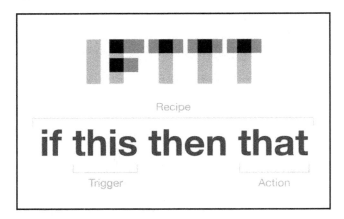

Enabling sensors with Particle Photon

In this section, we will use the Particle Photon to send a notification using IFTTT. When the motion sensor detects something, it will send a notification. First, we will have the following diagram for this project:

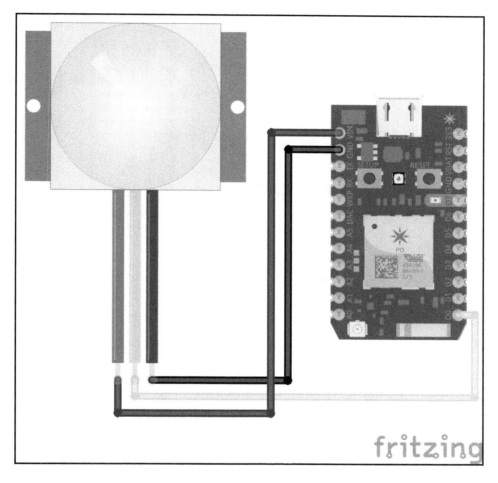

The `Particle.publish()` event will later act as an IFTTT trigger.

The code for the motion sensor is as follows:

```
#define PIR_PIN D0 // Replace D0 with the pin you used
#define MIN_TIME_BETWEEN_TRIGGERS 2000 // Time (in milliseconds) of no
motion before a new trigger can occur
void setup() {
   pinMode(PIR_PIN, INPUT);
}
void loop() {
   // PIR_PIN goes HIGH when motion is detected, stays HIGH for a few
seconds
   if (digitalRead(PIR_PIN)) {
       Particle.publish("motion-detected");
```

```
        // store current time in variable
        unsigned long motionTime = millis();
        // wait until no motion has been detected for
MIN_TIME_BETWEEN_TRIGGERS milliseconds before a new trigger can occur
        while(millis() - motionTime < MIN_TIME_BETWEEN_TRIGGERS) {
            if (digitalRead(PIR_PIN)) motionTime = millis();
        }
    }
}
```

Creating the IFTTT applet

Now in order to connect our devices and send notifications, we need to follow some important steps:

1. Create an IFTTT account (`https://ifttt.com/join`).
2. Connect a Particle to your IFTTT account (`https://ifttt.com/particle`).
3. Click on your username in the top-right corner of IFTTT's website and then click on **New Applet**:

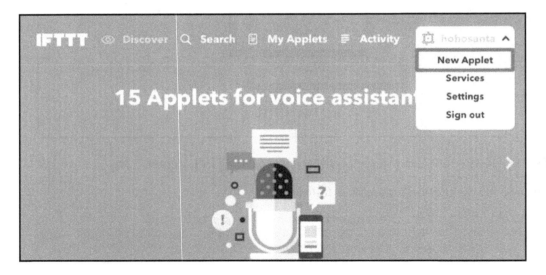

4. Click on **+this** in the **if +this then that** formula. Search for `particle` and select it:

5. Choose the option **New event published** for the trigger.

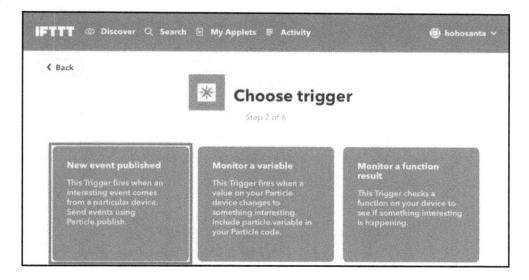

6. Type the **Event Name** to match the one in the Photon code. The event name is the text between the first set of quotes inside `Particle.publish("event-name")`.

Motion detector, this is the event that will occur when the notification is called by the condition, when the sensor detects movement, it is been called the function motion-detected and the device will send the notification to the Internet of things service:

```
Particle.publish("motion-detected");
```

Use `motion-detected` as the **Event Name**:

7. Finish your applet by adding an action. Choose from hundreds of services. Follow the prompts to finish creating your first applet. When you are finished, make sure the applet is turned on and ready for use.

In the last step, we need to choose the action service or notification:

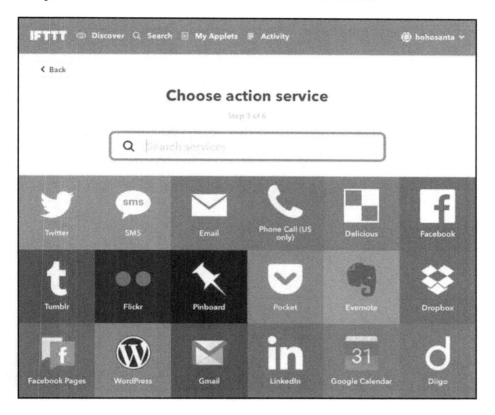

To finish, this is the last screen, and the slide needs to be **On**:

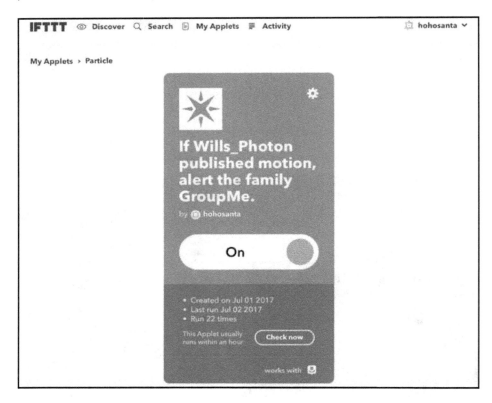

Sending notifications

In this section, we will see different notifications we can develop for the project we can use different buttons each of them can be clicked and send a different notification. Here we have some examples:

- Open the garage door
- Water the plants
- Open the main door
- Dim the lights
- Send an SMS when someone enters the house
- Send an SMS to turn the lights **On** or **Off**
- Make a phone call to order pizza

In the following figure, we describe the function of the WEB services configured in IFTTT, and this can help us send notifications to our devices through the internet:

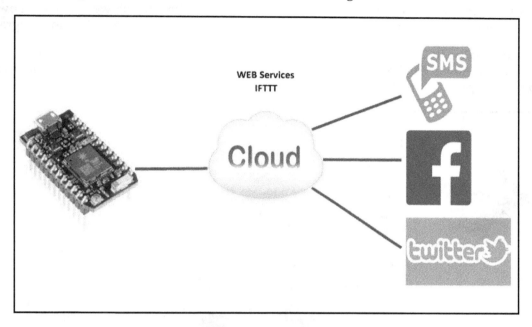

Now we have the part of the code that will have the notification actions:

The event name is the text between the first set of quotes inside `Particle.publish("event-name")`. Event Names from the examples are shown as follows:

- Order a pizza:

```
Particle.publish("order-pizza")
```

- Open the garage door:

```
Particle.publish("open-garage-door", "high");
```

- Make a phone call:

```
Particle.publish("make-a-phone-call");
```

Future ideas for projects

For future projects and when talking about IoT projects, in this section, we will show how we can develop an automatic system. To implement this, we will see the following diagram:

According to the following architecture, we will explain the steps that the system will perform and how it can be integrated into future ideas and projects using wireless devices and can be applied to real situations:

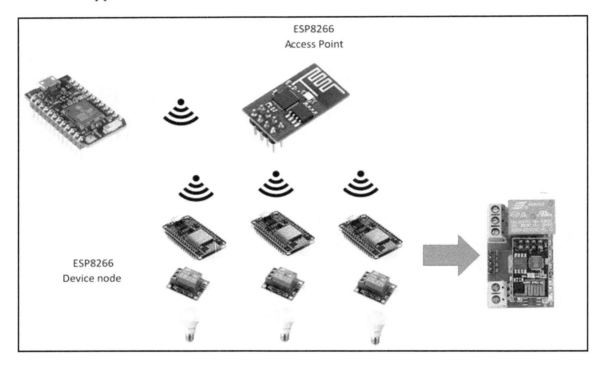

- The main brain of the system is still the Particle Photon; it connects to a wireless network.
- It connects to an access point built with an ESP8266, and it can be implemented with this feature. Some of the commercial products have a wireless module inside; in this case, we can develop our own device in order to connect other devices with a small device.
- The device nodes are configured as clients; in their pins, we configure and install a relay module, as shown in the figure. Then, we connect a bulb. These modules are configured like clients; they are connected to the access point.

- The complete system is controlled from the Particle Photon. It sends the signal, and it can be implemented. With some push buttons, each button will control each device connected to the client nodes.

- This link among ESP8266 devices is executed by IFTTT communication; the main device is connected to the access point.

- The system is connected to a WEB service; this is the bridge of communication among the devices.

- The access point with ESP8266 will help the system connect the final nodes that we want to control.

- To send notifications, we use IFTTT and register the devices that will be connected when the main brain sends the data, and the final nodes will receive the data when the service notifies them about the action.

- The relay module will be activated and the light bulbs will be on or off.

- If we want to control more electrical devices at home, we can place this module where we need it.

Summary

In this chapter, we built an interesting project for a home automation system that can be implemented at home. In the first part, we talked about the Particle Photon, how to set up the board, and some important characteristics of Particle Photon.

Then, we connected the board to its framework and looked at an example. After that, we created a project to have control over electrical devices or other devices at home. We created a smart button that can send notifications and have control over the things at home. The first part of the project was to decode the signals that we sent to the TV; for that, we used an Arduino UNO to decode. For the last part, we made a proposal to use, with the same architecture of the system, some devices such as ESP8266 to send and control our devices in a local network.

In the next chapter, we will build an interesting project using a smart camera and a Raspberry Pi board to detect face recognition and turn on a lock door.

6
Lock Down with a Windows IoT Face Recognition Door System

When we leave our homes and want to be sure that they are secure, we need to have control. If someone wants to get into the house, we need to take care of the things that we have inside; if a person rings the doorbell, we want to see who is ringing it. Or, we want to be able to detect whether somebody is near the house, see the face of the person, and control all these situations. We can control all the devices at home; we can reduce the number of robberies and monitor everywhere in the house. With a secure house, we will not worry about these things.

Home automation solutions are necessary for having a comfortable time at home. If we can control the entrance of the place, the number of people getting into the building, and the number of them that are leaving, this can grant permission to all the rooms at home or especially automate the lights of the rooms that belong to the person who detects in front of the webcam. Date and time can be stored in a database server in order to avoid the use of controlling the process manually; if we do this with an automatic security system, we can determine what is happening in the building.

In this chapter, we will build a project that will consist of a home security system that can control the entrance of a house using a face recognition interface, using a Raspberry Pi 3 and a webcam.

Specifically, we will cover the following topics:

- Getting started with the Raspberry Pi 3
- Installing and configuring Windows 10 IoT on the Raspberry Pi 3
- Installing Visual Studio Enterprise 2015
- Creating a first example
- Applications for the Internet of Things
- The architecture of the security system
- Future projects

Getting started: Installing and configuring Windows 10 IoT on the Raspberry Pi 3

You are almost ready to plug in the Raspberry Pi 3. Even though your hardware configuration is complete, you'll still need to complete the next section to power on the device. So let's figure out how to install an operating system.

In order to deploy and flash the image of Windows 10 IoT Core, we need to follow the steps outlined in the upcoming section.

Preparation

First, we need to install Windows 10 on a computer Desktop in order to flash Windows 10 IoT Core on the Raspberry Pi 3 device.

Installing Windows 10 Desktop

This is the link to get a Windows 10 trial version to install it on the PC `https://www.microsoft.com/en-us/software-download/windows10`.

Getting the package to install Windows 10 IoT Core

To get the package of the installation, we need to go to `https://www.raspberrypi.org/downloads/`:

1. First, we need to go to this link to get Windows 10 IoT Core `https://developer.microsoft.com/en-us/windows/iot/getstarted`.
2. Select the hardware of the device and the version of the operating system you will get for the device.

Downloading and installing Windows 10 IoT Core

In this part, we will need to download the Windows 10 IoT Core Dashboard from `http://go.microsoft.com/fwlink/?LinkID=708576`. After downloading the file, you need to open the dashboard application for Windows 10 IoT:

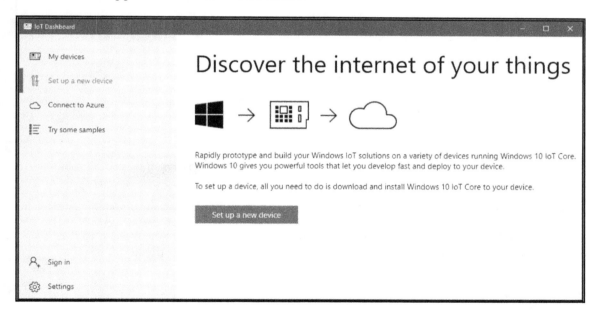

Deploying Windows 10 IoT Core on Raspberry Pi 3

In this part, we need to flash the image in our SD card.

Downloading and flashing the image

In the following screenshot, we show the steps to set up:

When we click on **Download and install** the image, the image will be copied into the SD card. In the following screenshot, we can see the process of copying the actual image:

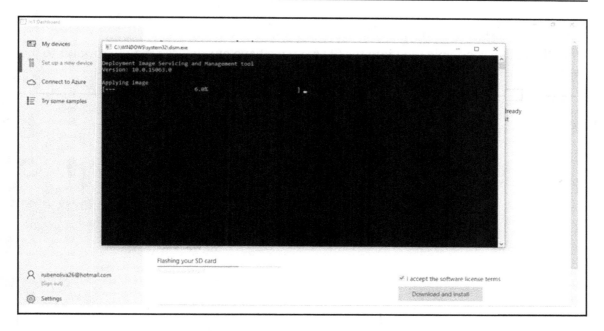

After this, we can see the files copied into the SD card:

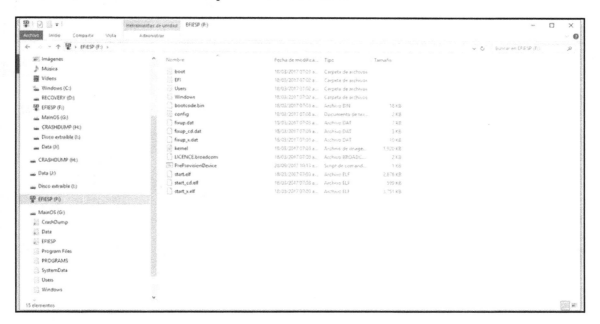

Connecting the board to the network

Once the image is downloaded into the SD card, we have our device ready to connect to the network:

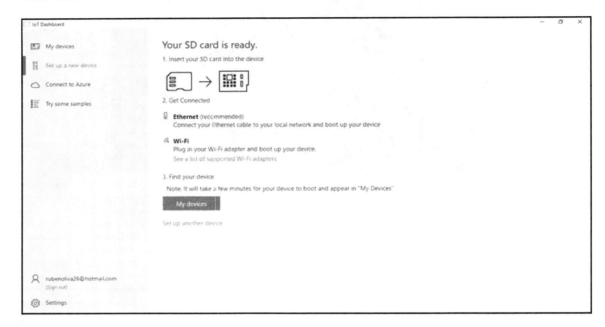

Attaching the MicroSD card on Raspberry Pi 3

When everything is ready, we will insert microSD card and have it ready to be deployed with the OS:

Putting them all together

Now that you are ready to boot your Raspberry Pi 3, plug in all the devices:

Turn on the power of your Raspberry Pi; you will see the first time that the board will boot:

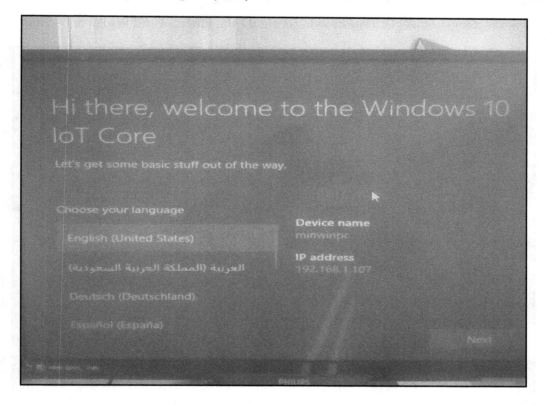

Ethernet and Wi-Fi connection

In the next screenshot, we will see how to configure the Ethernet and Wi-Fi connection:

If all is successful, you will see the final screen:

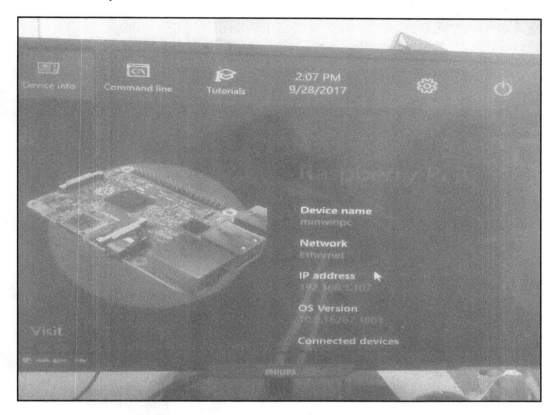

Now our device is ready to deploy and install Visual Studio, which we will look at in the next section.

Installing Visual Studio

When we install Visual Studio Enterprise 2015, for the installation process, check out **Universal Windows App Development Tools** include **Tools and Windows SDK**:

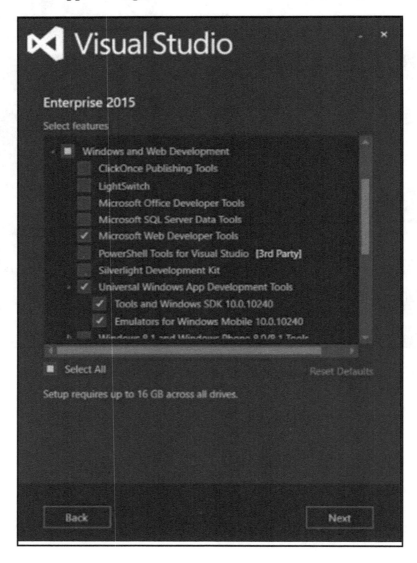

After installing Visual Studio and updating it, we will see the IDE of Visual Studio:

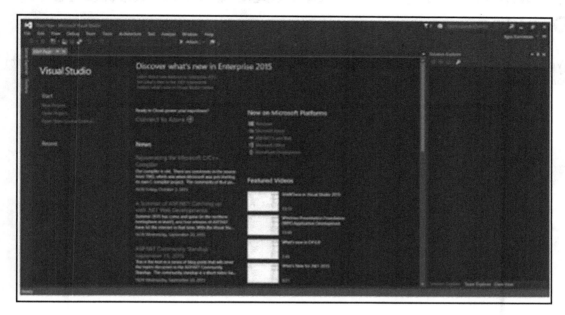

Enabling the Developer Mode on Windows 10 Desktop

We also need to enable the **Developer Mode** on Windows 10, so we can develop a program for Raspberry Pi 3:

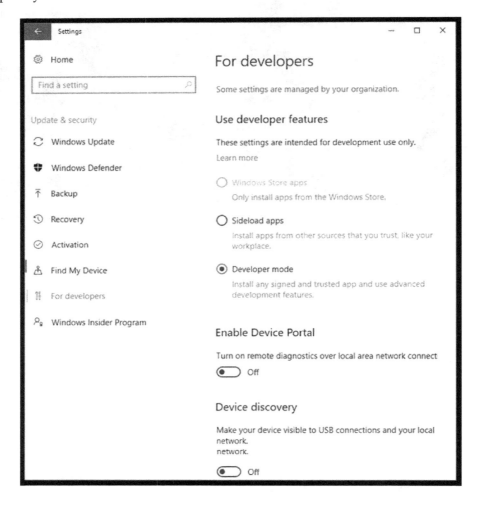

Go to the **Developer Mode**:

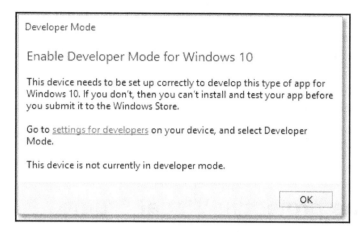

After this, we are ready to start a new program and develop an application.

Creating a first example

After completing the installation of Visual Studio and Windows IoT, we will move on to an example for testing our Raspberry Pi 3 and integrating everything we installed on the board. Let's take a look at this example:

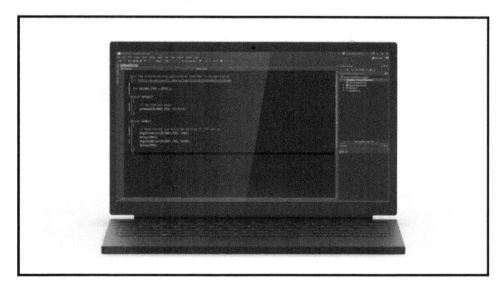

We need to follow these steps:

1. Install Microsoft Visual Studio; this step was performed in the previous section.
2. Install the Windows IoT project templates; go to `https://go.microsoft.com/fwlink/?linkid=847472`.

Writing our first application

We will create a simple blinking LED and connect an LED to your Windows 10 IoT Core device.

Loading the project in Visual Studio

You can find the source code by navigating to `samples/develop/`.

 Be aware that the GPIO APIs are only available on Windows 10 IoT Core, so this sample cannot run on your desktop.

Connecting the LED to your Windows IoT device

In order to make the connections, we will need some materials:

- An LED (any color)
- A 220 Ω resistor
- A breadboard and a couple of wires

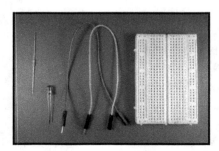

Hardware pins

In this figure, we can see the pins of the Raspberry Pi 3 (GPIOs):

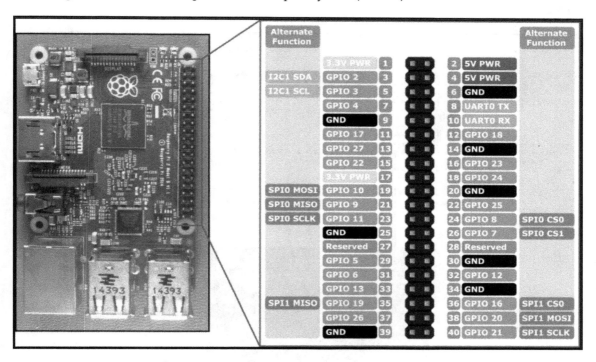

Hardware connections

In this section, we show the connections of the hardware connected with the LED to the Raspberry Pi 3:

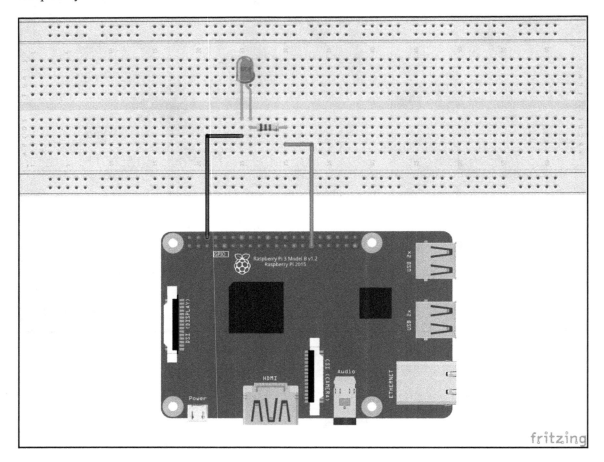

Deploying the app

First, you need to select the ARM architecture from the drop-down menu, select **Remote Machine**. This is the device that we will use to deploy the application:

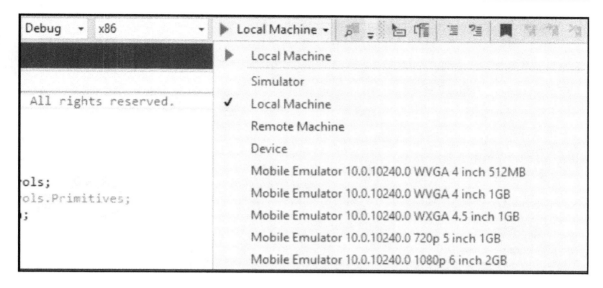

Here, we see the remote connections. It's recommended that you write the IP address of the Raspberry Pi:

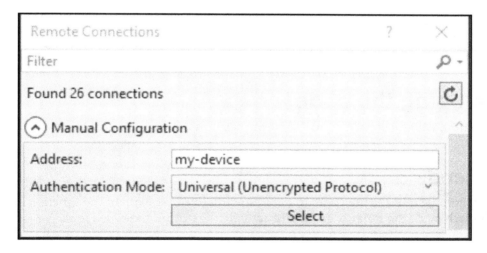

We can modify the properties by navigating to the explorer window:

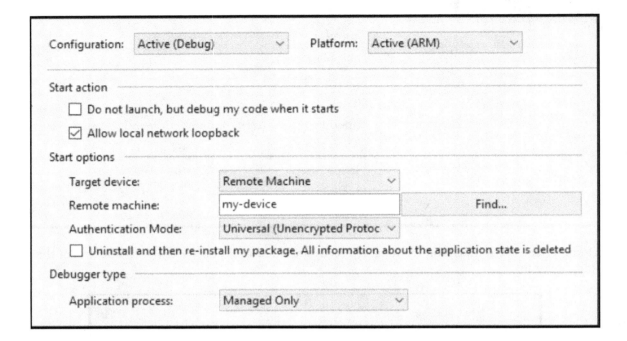

Developing the code

In this section, we will see the code for the application divided into different sections.

Timer code

In this section we will see the code. In the following lines we explain the function of the timer. In each cycle the timer execute and activate the output the time loops:

```
public MainPage(){
// ...
    timer = new DispatcherTimer();
    timer.Interval = TimeSpan.FromMilliseconds(500);
    timer.Tick += Timer_Tick;
    InitGPIO();
    if (pin != null)
    {
    timer.Start();
```

```
        }
    // ...
    }

    // This function will activate the GPIO, and it reads the state of the pin
    and it //compares the value of the pin if it's High, it assigns the value
    of LOW.

    private void Timer_Tick(object sender, object e)
        {
        if (pinValue == GpioPinValue.High)
        {
        pinValue = GpioPinValue.Low;
        pin.Write(pinValue);
        LED.Fill = redBrush;
        }
    else
        {
        pinValue = GpioPinValue.High;
        pin.Write(pinValue);
        LED.Fill = grayBrush;
        }
    }
```

Starting GPIO pins

In the following part of the code written in C#, we need to initialize the out-pins and manage them to control them:

```
using Windows.Devices.Gpio;
private void InitGPIO()
{
    var gpio = GpioController.GetDefault();
    // Show an error if there is no GPIO controller
        if (gpio == null) {
        pin = null;
        GpioStatus.Text = "There is no GPIO controller on this device.";
    return;
    }
    pin = gpio.OpenPin(LED_PIN);
    pinValue = GpioPinValue.High;
    pin.Write(pinValue);
    pin.SetDriveMode(GpioPinDriveMode.Output);
    GpioStatus.Text = "GPIO pin initialized correctly.";
    }
```

Modify the state of a pin

Once we have access to the `GpioOutputPin`, we can change the state of the LED; to turn the LED *on*, we just need to write the following:

```
pin.Write(GpioPinValue.Low);
```

If we want to turn the LED off, we need to write this:

```
pin.Write(GpioPinValue.High);
```

Running the application

In the following image, we have the execution screen of the application. The LED is on or off according to the time programmed in the application:

After connecting all the hardware, this is the final result:

Applications for the Internet of Things

In the next section we will show some applications for the Internet of Things.

Real-life examples of the Internet of Things

The Internet of Things is fascinatingly spread in our surroundings, and the best way to check it is to go to a shopping mall and turn on your Bluetooth. The devices you will see are merely a drop in the ocean of the Internet of Things. Cars, watches, printers, jackets, cameras, light bulbs, street lights, and other devices that were too simple earlier are now connected and continuously transferring data. Keep in mind that this progress in the Internet of Things is only 3-years old and it is not improbable to expect that the adaptation rate of this technology will be something that we have never seen before.

The increase in the number of internet users in the last decade has been exponential, reaching the first billion in 2005, the second in 2010, and the third in 2014. Currently, there are 3.4 billion internet users in the world. Although this trend looks unrealistic, the adaptation rate of the Internet of Things is even more excessive. The reports say that by 2020, there will be 50 billion connected devices in the world and 90% of the vehicles will be connected to the internet. This expansion will bring $19 trillion in profits by the same year. By the end of this year, wearables will become a $6 billion market with 171 million devices sold.

Smart home devices

The evolution of the Internet of Things is transforming the way we live our daily lives. People have already started using wearables and many other Internet of Things devices. The next big thing in the field of the Internet of Things is the **smart home**. Home automation or simply smart homes is a concept where we extend our home by including automated controls to things such as heating, ventilation, lighting, air-conditioning, and security. This concept is fully supported by the Internet of Things, which demands the connection of devices in an environment. Although the concept of smart home came to the surface in the 1990s, it hardly got any significant popularity among the masses. In the last decade, many smart home devices came onto the market due to major technology companies.

Wireless bulbs

The Philips Hue wireless bulb is another example of a smart home devices. It is a Bluetooth-connected light bulb that provides full control to the user through their cellphone. The bulbs can change to millions of colors and can be controlled remotely through the *away from home* feature. The lights are smart enough to sync with music:

Smart refrigerators

Home automation is incomplete without kitchen electronics, and Samsung has stepped into this race. The **Family Hub Refrigerator** is a smart fridge that lets you access the Internet and runs many applications. It is categorized as an Internet of Things devices as it is fully connected to the internet and provides various controls to the users:

This image belongs to Samsung.

Applications of the Raspberry Pi in the Internet of Things

With the usage of the Raspberry Pi 3, we can develop many projects for smart home automation homes. There are many smart-home devices that are in the market, but if we create our own it's very interesting and exciting. Using our Raspberry Pi 3 is a great choice for that. In this section, we will discuss the potential IoT projects you can implement using the Raspberry Pi 3.

The Raspberry Pi 3 Model B can be configured with various operating systems, but Raspberry Pi encourages the use of Raspbian, a Debian-based Linux operating system. Other operating systems compatible with the computer are as follows:

- Ubuntu Mate and Ubuntu Core (Snappy)
- Windows 10 IoT Core
- RISC OS
- Fedora
- Kali Linux
- Slackware
- Android Things

Media center using the Raspberry Pi 3

We can create our own media center at home with a Raspberry Pi 3. Configure the video streaming and with the audio configuration it's a great choice to do that it can also use the radio, Spotify, Pandora, Google Play Music, and other streaming platforms.

Cloud storage using Raspberry Pi 3

Google Drive, Microsoft OneDrive, Dropbox, and other services provide cloud storage, but usually, they are not free. We can configure and build our own cloud storage with a Raspberry Pi 3 and an external SD. Since the Model 3 B comes with built-in wireless LAN and Ethernet capabilities, we can make our local storage disk and can access from anywhere.

Tracker using Raspberry Pi 3

Using a Raspberry Pi 3 we can build a tracker with a GPS. It can be placed on any object that you want to monitor, for example we can monitor a bicycle and tracks all the road, it send data to a cloud server and store the route and displayed the information a map, also it can send a SMS message to a cell phone if the vehicle goes to other place or it stops

Web server using Raspberry Pi 3

Also we can set up a web server for publishing applications, the client can request a web page and the server answers with the page requested, some of the servers we can install is Apache web server instance on a Linux-based Raspberry Pi 3 is ideal for web developers.

We can also monitor the temperature and humidity of the conditions of the physical equipment and detects server's conditions.

Gateway for Bluetooth devices using Raspberry Pi 3

In our house we have many devices connected via Bluetooth, all the devices now need to connected to a gateway, and link them to the Internet of Things, We can build a gate way using a Raspberry Pi 3 and monitor the connected devices to the gateway. Nowadays the devices that we have at home are required and connected to the Internet of Things. The gateway will be the link among the devices connected with Bluetooth and the internet network.

In the next section, we will be building a lock down with a Windows IoT face recognition door system.

The architecture of the security system

In the following section, we will explain how the security system works and how it will be operated. The hardware connections of the system are as follows:

- The button (works as the doorbell of the system)
- A relay module that controls the lock or unlocks the door

- Connects the web camera that will detect the person that wants to enter the house
- Speakers that will notify the messages about the system
- When the button is pressed, it will request the image taken by the camera
- The Raspberry Pi 3 is connected to the internet by Ethernet interface

The following diagram shows the elements of the system:

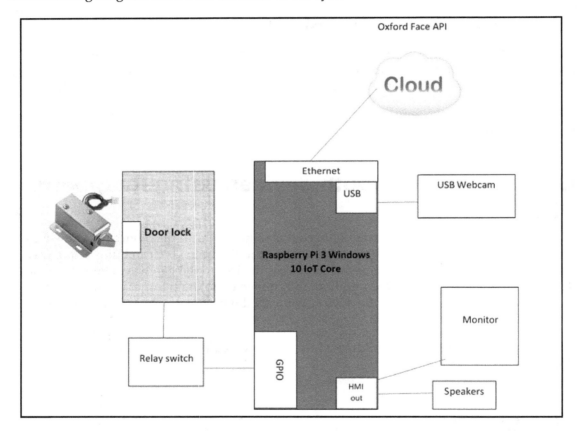

Materials required

These are materials that are required for this project:

- Raspberry Pi 3
- Perma-Proto breadboard half-size
- Leviton 12-Volt DC electric door strike
- SainSmart 2-channel 5V relay module
- Adafruit female DC power adapter
- Adafruit 12V—5A switching power supply
- Microsoft lifecam 3000 (recommended)
- Male-female jumper wires
- Female-female jumper wires
- Generic keyboard
- Generic mouse
- Generic speakers

Hardware connections

In the following diagram, we have the connections for the control and turning the lock *on* or *off*:

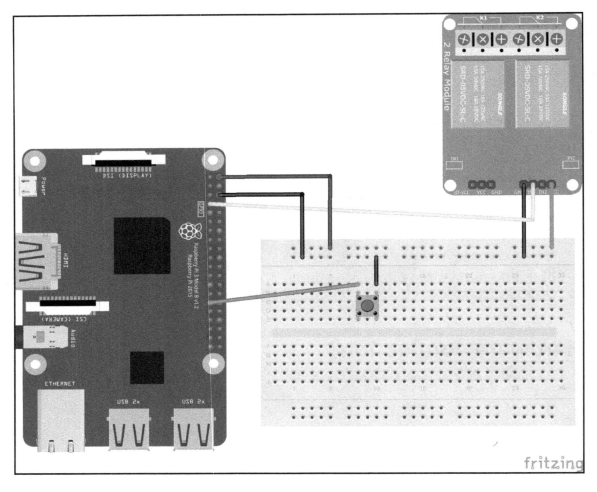

In the next diagram, we have the connections for the electrical part:

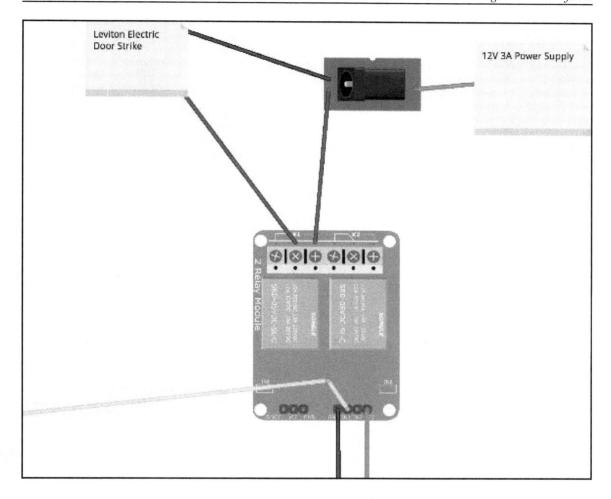

Initial setup

The following steps to create the project:

1. Set up a PC and the Raspberry Pi 3.
2. Next, wire the doorbell and power on the relay. The power relay will be used to lock and unlock the door.
3. Now wire the same power relay to the lock or the electric door strike.
4. Plug in your USB camera, keyboard, and mouse to the Raspberry Pi 3.

Software required

For this project, we need to have the following software and tools installed:

- Microsoft Visual Studio Enterprise 2015
- Microsoft Azure (facial recognition is done through Microsoft Face APIs within Project Oxford, hosted by Azure)
- Microsoft Windows 10 IoT Core

Software setup

Let's perform these steps:

1. Use Command Prompt to navigate to the path of the project:

   cd <your folder path>

2. Run the `git clone` command to download the project:

   git clone https://github.com/ms-iot/Facial-Recognition-Door.git

3. Open the `FacialRecognitionDoor.sln` solution file we just downloaded from Visual Studio.
4. On the right-hand panel, under the name of the project, open the `Constants.cs` file. You need to add the Oxford API key. To get the key go to `https://www.projectoxford.ai/doc/general/subscription-key-mgmt`.
5. Replace `OXFORD_KEY_HERE` with your new key:

```
/// <summary>
/// General constant variables
/// </summary>
public static class GeneralConstants
{
    // This variable should be set to false for devices, unlike the Raspberry Pi, that have GPU support
    public const bool DisableLiveCameraFeed = true;

    // Oxford Face API Primary should be entered here
    // You can obtain a subscription key for Face API by following the instructions here: https://www.projectoxford.ai/doc/
    public const string OxfordAPIKey = "OXFORD_KEY_HERE";    ←

    // Name of the folder in which all Whitelist data is stored
    public const string WhiteListFolderName = "Facial Recognition Door Whitelist";

}
```

This is the code got it from OXFORD credentials:

```
public const string OxfordAPIKey = "OXFORD_KEY_HERE";
```

Software development

In this section, we have the main code of the project written in C#:

```
namespace FacialRecognitionDoor
{
/// <summary>
/// General constant variables
/// </summary>
public static class GeneralConstants
{
// This variable should be set to false for devices, unlike the Raspberry
Pi, that have GPU support
public const bool DisableLiveCameraFeed = true;

// Oxford Face API Primary should be entered here
// You can obtain a subscription key for Face API by following the
instructions here:
https://azure.microsoft.com/en-us/try/cognitive-services/
public const string OxfordAPIKey = "OXFORD_KEY_HERE";

// Enter the API endpoint address.
// If you have a 'free trial' key, you can find the here:
https://azure.microsoft.com/en-us/try/cognitive-services/my-apis/
// If you have a key from Azure, find your account here:
https://portal.azure.com/#blade/HubsExtension/Resources/resourceType/Micros
oft.CognitiveServices%2Faccounts.
public const string FaceAPIEndpoint =
"https://westus.api.cognitive.microsoft.com/face/v1.0";

// Name of the folder in which all Whitelist data is stored
public const string WhiteListFolderName = "Facial Recognition Door
Whitelist";
}

/// <summary>
/// Constant variables that hold messages to be read via the SpeechHelper
class
/// </summary>
public static class SpeechContants
{
public const string InitialGreetingMessage = "Welcome to the Facial
```

```
Recognition Door! Speech has been initialized.";

public const string VisitorNotRecognizedMessage = "Sorry! I don't recognize
you, so I cannot open the door.";
public const string NoCameraMessage = "Sorry! It seems like your camera has
not been fully initialized.";

public static string GeneralGreetigMessage(string visitorName)
 {
 return "Welcome to the Facial Recognition Door " + visitorName + "! I will
open the door for you.";
 }
 }

/// <summary>
/// Constant variables that hold values used to interact with device Gpio
/// </summary>
public static class GpioConstants
{
// The GPIO pin that the doorbell button is attached to
public const int ButtonPinID = 5;

// The GPIO pin that the door lock is attached to
 public const int DoorLockPinID = 4;

// The amount of time in seconds that the door will remain unlocked for
 public const int DoorLockOpenDurationSeconds = 10;
 }
}
```

Running the interface

After deployment, the application we can see the following screenshots:

We can capture a photo:

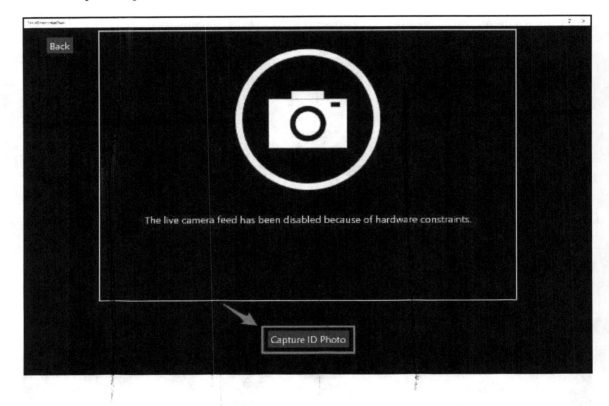

Confirm the captured selfie and type in the name of the person:

After this, we can see the image captured in the left-hand corner:

Now we're ready to unlock the door. Position the registered user in front of the webcam and press either the virtual door icon located next to the previously used *plus* icon or the physical **doorbell** button you wired up. You should hear audio feedback informing you that the door has been unlocked!

Now try pressing the doorbell button when an unregistered user is in front of the door. You should hear audio feedback informing you that the door has detected a stranger, and it will not unlock!

Integrating the system and putting it all together

In the following image, we have the complete prototype of the system:

Future ideas

For future ideas and improving this project, we will discuss ideas about how to develop a real situation. In the following figure, we have the application of an entrance to a school, which can do following:

- Control the entrance with the face recognition system using a webcam
- The entrance door can be configured with a database that logs the date and time a person arrives
- I can store the data about the have the number of people that have entered or left the building
- In some places, it requires the name of the person, the department that he/she visits the place they came from, and so on
- All this information can be stored in a database server and can be controlled from the internet
- When the person or the student enters or leaves the system will send an SMS to their parents

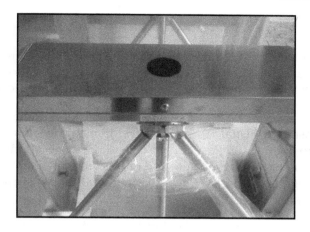

In the following figure, we have the architecture of the system, which will send the SMS when somebody enters:

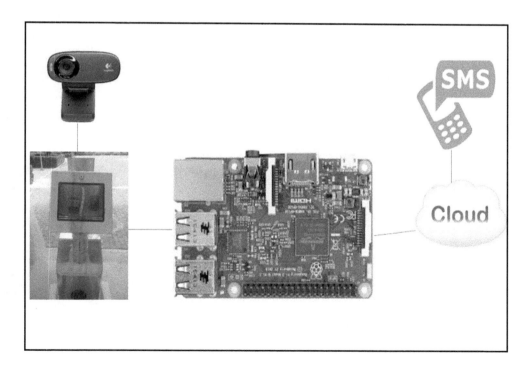

Summary

In this chapter, we built a security system that can be applied to different applications. We used a Raspberry Pi 3 as a central interface in the beginning of the project. We set it up, installing Windows 10 IoT Core and configuring and installing the required software and hardware.

In the next part, we installed Visual Studio for developing applications. You learned about the necessary things to operate the Raspberry Pi 3 and the applications that can be developed for the Internet of Things. At the end of this chapter, we built a lock down with a Windows IoT face recognition door system and integrated the software and hardware required for this interesting project.

For future projects, we can apply this kind of technology for doing these kinds of developments and be assured that our devices are really putting forward interesting ideas and can be used in the real world.

Index